LAB WORKBOOK

Animal Science

Principles of Production, Management, and Selection

Jessica N. Fife • Noe A. Gomez

Publisher
The Goodheart-Willcox Company, Inc.
Tinley Park, IL
www.g-w.com

ISBN 979-8-89118-198-4

3 4 5 6 7 8 9 – 25 – 28 27 26

Image Credits. Front cover: SteveOehlenschlager/iStock via Getty Images; Dr. Jeremy Falk, University of Idaho; Lindsay Hanewich Photography; yod 67/Shutterstock.com

Introduction

This workbook is designed for use with *Animal Science: Principles of Production, Management, and Selection*. The chapters in the workbook correspond to those in the text and should be completed after reading the appropriate chapter.

Each chapter of the workbook contains review questions related to the text chapter to enhance your understanding of the principles presented. These questions include matching, true or false, multiple choice, fill in the blank, and short answer.

Reading *Animal Science: Principles of Production, Management, and Selection* and using this workbook will help students master the principles of livestock and domestic animal production and care as presented.

The workbook chapters also contain activities related to textbook chapter content. The activities range from chapter content reinforcement to real-world application. It is important in these activities to understand any safety procedures set forth by your teacher.

Contents

Name ______________________________ Date ______________ Period __________

CHAPTER 1 Animals and Society

Chapter Review

Carefully study the chapter and answer the following questions.

1. _____ When were horses domesticated?
   A. 500 years ago
   B. 4500 BCE
   C. 3000 BCE
   D. Between 7000 and 6000 BCE

2. Explain domestication and its contributions to animal agriculture.

____

____

3. _____ work specifically on _____ of science to improve the health, welfare, and _____ of animals that are used by humans.

____

____

____

4. Describe the scope of the animal industry in the United States and globally.

____

____

5. What was the first animal to be domesticated for food and when was it domesticated?

____

____

6. _____ When were cattle, poultry, and pigs domesticated?
   A. 500 years ago
   B. 4500 BCE
   C. 3000 BCE
   D. Between 7000 and 6000 BCE

7. Describe how animals are classified.

____

8. _____ is the term for animals that come from a domesticated species but live in the wild with no human assistance.

____

9. _____ is the Swedish botanist who developed the seven-level classification system we use to classify living things.

____

10. Each species of animal has different needs related to _____, _____, and _____.

____

____

____

11. What feral animal is a major issue in the United States?

____

12. Explain the different uses for animals in society.

____

13. _____ Animal products account for _____of the value of US agricultural products, and often exceed _____ in sales per year.
    A. Less than half; $100 billion
    B. More than half; $100 billion
    C. Less than half; $10 billion
    D. More than half; $10 billion

14. List the five ways an animal must be different from the wild species to be considered domesticated.

____

____

____

____

____

Key Terms

Carefully study the chapter and then use the word bank to match the term to the correct definition.

A. zoology	D. binomial nomenclature	G. animal science
B. domestication	E. taxonomy	H. animal industry
C. service	F. by-products	I. product

1. _____ Something that is refined for sale
2. _____ The scientific classification of animals
3. _____ The sector of agriculture that includes all the work related to raising animals, processing animal products, and marketing the goods and services from animals
4. _____ The scientific study of the behavior, structure, physiology, classification, and distribution of animals
5. _____ Products outside of food and fiber that come from animals stemming from a desire to use as much of the animal as possible after it is harvested
6. _____ A subcategory of zoology which includes the care and management of animals that live and work with people
7. _____ The custom of referring to an animal by its genus and species
8. _____ The process of changing a wild animal for the benefit of humans
9. _____ Any type of work completed by an animal for human benefit

Name ______________________________

Critical Thinking

Using the information learned in this chapter, read each question and provide a detailed answer.

1. Reflect on what you've learned in this chapter about animals and their impact on our lives as a means for food, companionship, travel, and labor. How would your daily life be different without animals? Do you think you would see a difference?

2. Now consider how humanity's existence would be different without animals. What do you think our lives would be like if we didn't have domesticated animals for companionship, food, or as a means for travel and labor throughout history? Could we live without them?

Name ______________________________ Date ______________ Period __________

Activity: Animals and Society

Animal Products in Everyday Life

Required Materials

- Lab notebook
- Computer with internet access
- Google Slides/Powerpoint™

Activity Specification

Pick a day to carefully observe animals and animal products as you 1) are on your way to school, 2) are in school, and 3) head home from school. Note everything from the ants crawling on your desk at school, to your neighbors walking their pets, to the turkey sandwich you may eat for lunch! Make a detailed list in your lab notebook of all the animals and animal products that you observe.

If you cannot account for at least eight observations in a single day, think about eight animal interactions you have had over the past year, or ask your instructor for assistance.

Summarize your findings in a five- to eight-slide presentation (not including a title slide). You will deliver a four-minute presentation where you share your observations with the class.

Reflective Questions

Answer the following questions relating to the activity.

1. Analyze the presentations of your peers and explain at least two common themes between your animal observations and those of your classmates.

____

____

____

2. Evaluate a positive aspect/skill of your peers' presentations that stood out to you. What is this specific aspect/ skill and how you can incorporate it into future presentations?

____

____

____

3. Create a follow-up question regarding an area of interest you may have about animals or animal products following the class presentations on this topic. What might you be curious to learn more about?

____

____

____

Name ________________ Date ________________ Period ____________

CHAPTER 2 Careers in the Animal Industry

Chapter Review

Carefully study the chapter and answer the following questions.

1. _____ Which of the following are the main hazards of working with animals? Select all that apply.
   A. Hunting
   B. Handling live animals
   C. Machinery and equipment
   D. Hazardous materials

2. Your _____, _____, and _____ will influence which types of careers you might enjoy.

3. Accidents while operating machinery and equipment are more likely to occur when the operator is _____, _____, or _____.

4. _____ Livestock handling is one of the leading causes of injury on farms and ranches, accounting for up to _____ of all agricultural injuries.
   A. 75%
   B. 24%
   C. 97%
   D. 42%

5. According to the Food and Agriculture Organization (FAO) of the United Nations, almost _____ people work in the animal science industry worldwide.

6. _____ Which of the following is *not* a characteristic of a good employee?
   A. Energy
   B. Self-awareness
   C. Never asks questions
   D. Curiosity

7. Describe the difference between pathogens and zoonosis.

8. It is estimated that more than _____ new animal science jobs are available each year in the United States, but there are only about _____ qualified candidates to fill those positions.

9. Describe three ways to prevent contracting a zoonotic disease.

____

____

____

10. List and provide a brief explanation of the job application's four steps.

____

____

____

____

11. List the five main career pathways in animal science.

____

____

____

____

____

Key Terms

Carefully study the chapter and then use the word bank to match the term to the correct definition.

A. allergen
B. Safety Data Sheet (SDS)
C. résumé
D. pathogen
E. cover letter
F. zoonosis
G. animal health careers
H. production and management careers
I. education and outreach careers
J. scientific research careers

1. _____ A substance causing an auto-immune response
2. _____ A document meant to give the employer information about your interest in the position and additional information about your qualifications
3. _____ Careers that involve breeding, feeding, raising, and caring for animals as a part of the production cycle
4. _____ Disease that can be transmitted between animals and humans
5. _____ Careers that involve providing a health-related service to animal owners
6. _____ Bacteria, viruses, or microorganisms that can cause disease
7. _____ Careers that involve conducting scientific examinations of animal-related topics
8. _____ A brief overview of your education, qualifications, and skills
9. _____ Careers that involve teaching people involved with the animal science industry about production practices and emerging technologies
10. _____ Includes information such as the properties of a chemical; the physical, health, and environmental health hazards posed by the chemical; protective measures to be taken in working with the chemical; and safety precautions for handling, storing, and transporting the chemical

Name ______________________________

Critical Thinking

Using the information learned in this chapter, read each question and provide a detailed answer.

1. Think of a subject area you are interested in—maybe it is science, English, or math. Now, consider how that subject matter might be used within animal agriculture careers and record your ideas below.

2. Do any of these sound like a career you might be interested in? Does it surprise you to see how other school subjects connect to animal agriculture? Why or why not?

Name ______________________ Date ______________ Period __________

Activity: Careers in the Animal Industry

Working in Animal Science

Required Materials

- Lab notebook
- Pen or pencil

Activity Specification

Your instructor will bring a guest speaker into the classroom today. As the guest speaker presents, jot down three higher-order questions that you can ask the speaker when they finish their presentation. In a higher-order question, you ask the speaker about how they critique, check, generate, plan, or produce in their daily work tasks. (Example for a veterinarian: How do you quickly change tracks when a surgery goes wrong?)

Question 1: ______________________

Answer: ______________________

Question 2: ______________________

Answer: ______________________

Question 3: ______________________

Answer: ______________________

Name ______________________________

Reflective Questions

Answer the following questions relating to the activity.

1. Analyze the guest speaker's attitude toward their job. Did their responses encourage or discourage you from pursuing this career? Why?

____

____

____

2. Evaluate the answers to your questions. What stands out to you about the speaker's daily work routine?

____

____

____

3. Create a title and a brief description of a job that you would be interested in applying for in the future.

____

____

____

Notes

Name ______________________ Date ______________ Period ________

CHAPTER 3 Anatomy and Physiology

Chapter Review

Carefully study the chapter and answer the following questions.

1. Identify and describe the levels of cellular organization from largest to smallest.

____

____

____

____

____

2. _____ What term is used to describe the many small parts of an animal cell that have specific functions?
   A. Vacuoles
   B. Tissues
   C. Organisms
   D. Organelles

3. Identify and differentiate between the four types of animal tissues.

_____________________ _____________________

_____________________ _____________________

4. _____ The smallest blood vessels are called_____.
   A. nares
   B. trachea
   C. alveoli
   D. capillaries

5. Explain the functions of the following animal systems.

A. Skeletal _______________________________

B. Muscular _______________________________

C. Reproductive _______________________________

D. Digestive _______________________________

E. Circulatory _______________________________

F. Urinary _______________________________

G. Respiratory _______________________________

H. Nervous _______________________________

I. Immune _______________________________

J. Endocrine _______________________________

K. Integumentary _______________________________

6. How do cells differ between organ systems?

____

____

7. _____ Which of the following is *not* one of the four stomach chambers of a ruminant animal?
   A. Proventriculus
   B. Reticulum
   C. Omasum
   D. Rumen

8. _____ The _____ is a temporary storage compartment where food can be held until the avian digestive system is ready to process it.
   A. proventriculus
   B. reticulum
   C. crop
   D. gizzard

9. _____ There are two main types of nerves that work to send signals throughout the body: _____ and _____.
   A. dendrites; axons
   B. afferents; efferents
   C. dendrites; efferents
   D. axons; afferents

10. What five qualities classify an organism as living?

______________________________ ______________________________

______________________________ ______________________________

____

Key Terms

Carefully study the chapter and then use the word bank to match the term to the correct definition.

A. omasum
B. ruminant digestive system
C. mitosis
D. modified monogastric digestive system
E. abomasum
F. reticulum
G. rumen
H. physiology
I. pharynx
J. monogastric digestive system

1. _____ Stomach chamber #2; a large fermentation area filled with microbes that digest cellulose
2. _____ The process used to replicate cells for growth
3. _____ Stomach chamber #3; this is where digested food is pressed between many layers of tissue to remove excess water
4. _____ A simple digestive tract that contains a stomach with only one chamber
5. _____ A digestive tract that contains a stomach with multiple specialized compartments designed to efficiently break down cellulose
6. _____ Stomach chamber #1; sometimes called the honeycomb or hardware stomach; sorts feed based on consistency and catches foreign materials ingested by the animal
7. _____ The section of the trachea that is shared with the digestive system in some animals
8. _____ The study of how organisms function
9. _____ A simple digestive tract that contains a stomach with only one chamber but includes a functional cecum
10. _____ Stomach chamber #4; section where digestive juices are added to the feed; also called the true stomach

Name ____

Critical Thinking

Using the information learned in this chapter, read each question and provide a detailed answer.

1. Draw a map of the ruminant stomach and, in your own words, label and briefly explain the job and importance of each chamber in the correct order.

2. What types of animals are ruminants? How does the ruminant digestive system help them to survive and thrive?

Name ______________________ Date ______________ Period ________

Activity: Anatomy and Physiology

Engineering Livestock Skeletons

Required Materials

- 8.5" × 11" art paper
- Thin-line markers
- Scissors
- Quick-dry glue
- 30 toothpicks

Activity Specification

Choose a livestock species (bovine, equine, ovine, etc.). Using toothpicks (which can be cut into various sizes) and glue, create a skeletal outline of the animal. Clearly label the following bones and locations: skull, mandible, scapula, humerus, sternum, ulna, radius, carpus, metacarpus, phalanges, ribs, patella, tibia, fibula, tarsus, femur, pelvis, cervical vertebrae, thoracic vertebrae, lumbar vertebrae, sacral vertebrae, and coccygeal vertebrae.

Reflective Questions

Answer the following questions relating to the activity.

1. Analyze the structure of the skeletal model. Which part of this skeleton do you think is most prone to bone injuries? Why?

____

____

____

2. Evaluate the skeletal model you created. Why is there so much skeletal complexity centered around the skull and thoracic areas?

____

____

____

Name ____

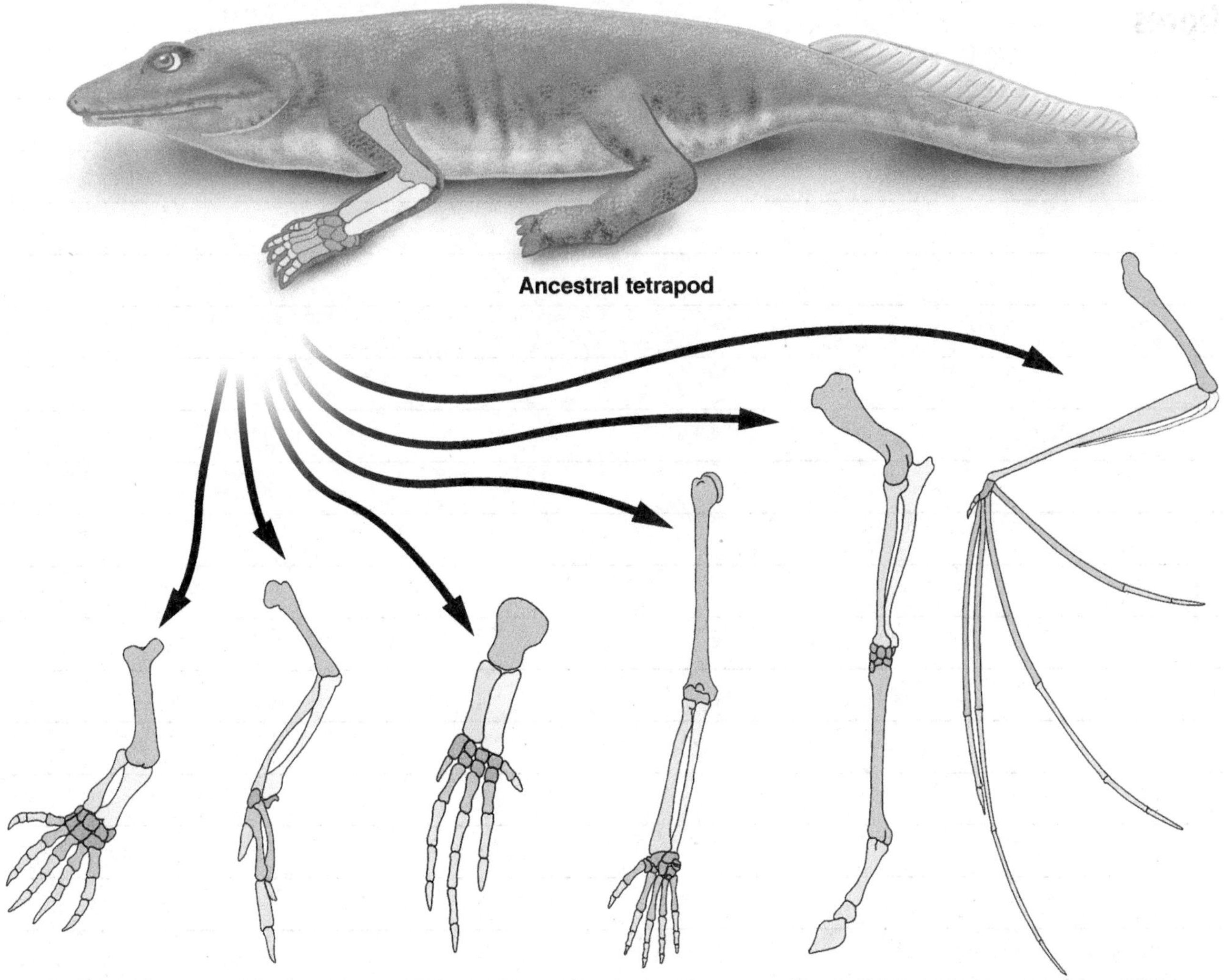

Amadeu Blasco/Shutterstock.com

3. Using the image above, create an explanation for an elementary-aged student about why it is important that we learn about different skeletal systems to improve our understanding of both animal and human health.

____

____

____

Notes

Name ______________________ Date ____________ Period ________

CHAPTER 4 Maintaining Animal Health

Chapter Review

Carefully study the chapter and answer the following questions.

1. List the main causes of illness in animals.

____

____

____

____

____

2. Describe how nutrition is related to illness, and how nutrition-related illnesses may be treated.

____

____

3. _____ cause illness by getting into the body and reproducing, interrupting body systems and functions.

____

4. Describe how genetic disorders are related to illness, and how genetic-related illnesses may be treated.

____

____

5. _____ cause illness by invading the body and taking over cell reproduction functions to replicate.

____

6. _____ Bovine Spongiform Encephalopathy (BSE) is an example of which of the following types of illness?
   A. Fungal
   B. Prion-based
   C. Viral
   D. Bacterial

7. Describe vital signs and how they indicate potential health issues.

____

____

8. _____ cause illness by attaching themselves to healthy animals.

____

9. _____ What are mycoses an example of?
   A. Bacterial infection
   B. Fungal infection
   C. Viral infection
   D. Parasite

10. What is important for producers to do each day to recognize observable indicators of declining animal health?

____

____

11. Describe biosecurity and its relationship with animal health.

____

____

12. List five ways producers keep track of animal health records.

____

____

____

____

____

13. Identify the three main components of radio frequency identification (RFID).

____

____

____

14. _____ What causes mange?
    A. Mites
    B. Ticks
    C. Coccidia
    D. Scrapie

Key Terms

Carefully study the chapter and then use the word bank to match the term to the correct definition.

A. ear notching	E. prions	I. subcutaneous injection
B. elastrator	F. gavagel	J. withdrawal time
C. antibodies	G. intramuscular injection	
D. castration	H. intravenous injection	

1. _____ A tool that expands thick rubber bands to be placed over a portion of an animal's body to cut off the blood supply
2. _____ Injection given directly into the vein of an animal
3. _____ The administration of medication through a tube passed directly into the stomach
4. _____ Injection absorbed into the body through the muscle
5. _____ Protein particles that infect living cells and can trigger the proteins present in an animal's brain to fold abnormally
6. _____ A process that removes small sections of ears corresponding to animal identification numbers
7. _____ The amount of time following medication administration before an animal can enter the food supply
8. _____ Specialized cells that act like a set of instructions for fighting a recently defeated pathogen in the future
9. _____ The removal of the testicles of livestock animals
10. _____ Injection given just beneath the skin in the animal's neck or where skin is loose

Name ______________________________

Critical Thinking

Using the information learned in this chapter, read each question and provide a detailed answer.

1. Practices like castration or ear notching can seem harmful or cruel. Are they? Describe these practices and explain how and why they are used to maintain animal health.

____

____

____

____

____

____

2. Select a cause of illness (bacteria, virus, parasite, or fungi), and give a detailed explanation of how it causes illness, how that illness may be treated, and two or three examples of that type of illness.

____

____

____

____

____

____

Name ______________________ Date ____________ Period ________

Activity: Maintaining Animal Health

Practicing Injection Technique

Required Materials

- Syringe
- Capped needle
- Biohazard container
- Strip of leather
- ½ chicken breast
- Unripe pear
- Beaker
- Water
- Green food dye
- PPE (goggles, gloves, lab coat)

Activity Specification

Today you will practice injection technique. Chicken breast simulates puncturing the skin of animals with softer tissue, such as cats and dogs. Placing leather over chicken breast can simulate puncturing the skin of animals with a thicker hide, such as cattle. An unripe pear simulates the rigidity of puncturing tissue that is tense, such as the neck of a large boar.

Types of Injections

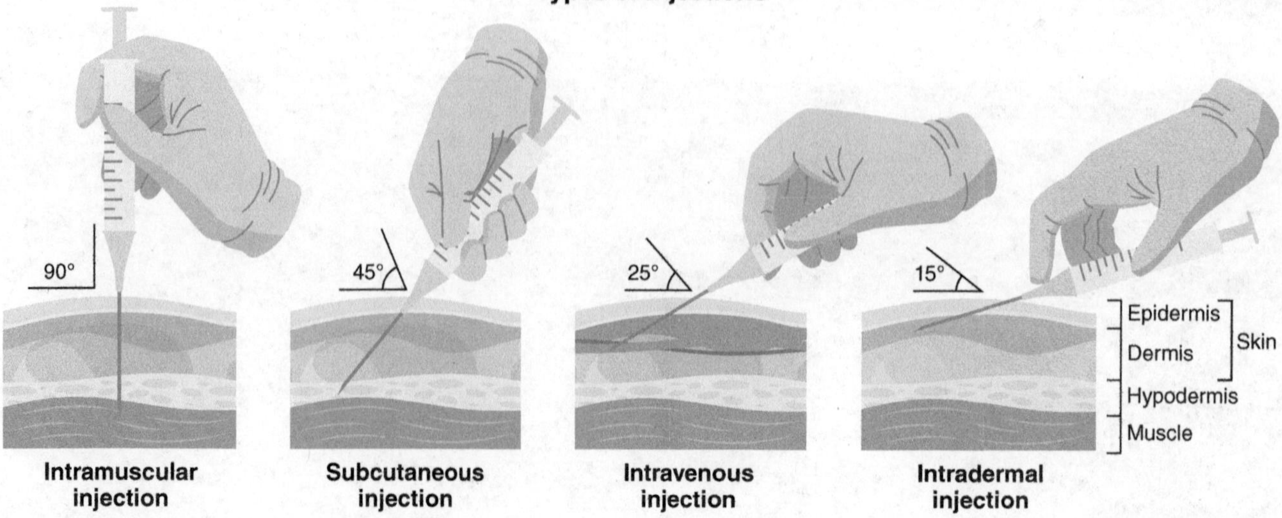

inspiring team/Shutterstock.com

Your instructor will ask you to demonstrate mastery of delivering the following types of injections at the appropriate angles:

Intramuscular injections, at 90° from site of injection

Subcutaneous injections, at 45° from site of injection

Intravenous injections, at 25° from site of injection

Intradermal injections, at 15° from site of injection

Name ______________________________

Reflective Questions

Answer the following questions relating to the activity.

1. Analyze the structure of a syringe. In the space provided below, design a syringe gun that can be used to inject an animal with a thick hide, such as a rhinoceros (search "syringe gun" online to see examples). You must label and include the following parts: plunger, barrel, needle hub, and needle. Label any additional parts as needed.

2. Evaluate the four different types of injections. Describe an application of each.

____

____

____

3. Create a simple three- to four-sentence safety description for employees in a veterinary clinic regarding the proper handling and disposal of syringes.

____

____

____

____

Notes

Name ______________________ Date ____________ Period ________

CHAPTER 5 Feeds and Feeding

Chapter Review

Carefully study the chapter and answer the following questions.

1. Ruminant animals can consume diets _____ in roughages, due to their more complicated digestive systems.

____

2. List the factors that determine the amount of each type of nutrient an animal requires.

____

____

____

____

____

____

____

3. What types of livestock animals thrive on a high-roughage diet?

______________________ ______________________

______________________ ______________________

4. _____ Which of the following is high in energy and total digestible nutrients (TDN) and low in fiber content?
   A. Carbohydrates
   B. Concentrates
   C. Cellulose
   D. Carboxyl groups

5. Describe medicated feed, its purpose, and special steps that are taken when it is used.

____

____

____

6. _____ There are _____ recognized vitamins that contribute to animal nutrition.
   A. 10
   B. 11
   C. 12
   D. 13

7. The types of bonds in the fat molecule determine whether fat is considered _____ or _____.

____

____

8. _____ are organic compounds that are essential for life.

____

9. ______ are inorganic substances, meaning they do not contain _____.

____

____

10. _____ Roughages contain more than _____ fiber when dried and are tough to digest.
   A. 20%
   B. 13%
   C. 18%
   D. 75%

11. Describe TDN and how it is measured.

____

____

Key Terms

Carefully study the chapter and then use the word bank to match the term to the correct definition.

A. amino acid
B. minerals
C. disaccharide
D. carboxyl group
E. crude protein
F. legumes
G. palatable
H. roughage
I. forage
J. silage

1. _____ Inorganic substances required by the body to perform life functions
2. _____ A feed that is low in energy and TDN and high in fiber content
3. _____ Appealing to eat
4. _____ Molecules made of carbon, hydrogen, oxygen, and nitrogen atoms, along with the same carboxyl group found in fats
5. _____ A plant grown to be fed to grazing animals
6. _____ A type of carbohydrate that is formed by combining two simpler carbohydrates
7. _____ The amount of protein based on the nitrogen content of the feed
8. _____ A molecular chain containing one carbon atom, two oxygen atoms, and one hydrogen atom (COOH)
9. _____ Plants that have a symbiotic relationship with bacteria, allowing them to take nitrogen from the soil and store it as protein
10. _____ A forage or grain crop that is cut and stored in anaerobic conditions

Critical Thinking

Using the information learned in this chapter, read the question and provide a detailed answer.

1. Compare and contrast the nutritional needs of livestock animals and people.

____

____

____

____

____

____

Name ______________________________ Date ______________ Period __________

Activity: Feeds and Feeding

Constructing a Particle Separator

Required Materials

- Computer with internet access
- Mixed livestock feed ration
- Popsicle sticks
- Thick 8.5" × 11" art paper
- Glue

Activity Specification

Using the internet, go to the Pennsylvania State University Extension website. Search for the Pennsylvania State University Particle Separator and research its function and purpose. Onsite separation of livestock total mixed rations (TMR) is essential to ensure that a producer can identify the relative percentage of particle size that they are feeding their animals. A balanced particle size ratio is essential at each stage of an animal's life to ensure digestive health.

Using the provided popsicle sticks, construction paper, and glue, construct a particle separator that can separate feed particles into at least three different sizes.

Reflective Questions

Answer the following questions relating to the activity.

1. Analyze the success of your particle separator. What can be done to promote a more representative collection of feed particles?

____

____

____

2. Evaluate the TMR ration based on its unseparated appearance. What textures and smells are most prominent?

____

____

____

3. Explain the purpose of a TMR particle separator to someone with little knowledge of animal feeds and feeding.

____

____

____

Notes

Name ______________________ Date ____________ Period ________

CHAPTER 6 Genetics

Chapter Review

Carefully study the chapter and answer the following questions.

1. Describe the importance of genetic improvement and give an example of how genetic improvements to animals have been successful based on what you learned in this chapter.

____

____

____

2. _____ What determines the traits that are shown in an EPD report?
   A. Genetic panel test results
   B. Purpose of the animal
   C. Breed registration status
   D. Punnett square

3. DNA in each cell is housed in the nucleus. When the cell needs to carry out an action, a copy of the instructions is made through a process called _____.

____

4. _____ is the Latin word for *same*, while _____ refers to one half of the genetic material.

____

____

5. What is the name for the female parent of an animal?

____

6. _____ Which of the following is shown as the phenotype if it is present in the allele pair?
   A. Dominant traits
   B. Recessive traits
   C. Incomplete dominance
   D. Heterozygous traits

7. Describe the purpose of a Punnett square and briefly explain how it is used.

____

8. RNA has only one strand, which is made up of _____ _____, rather than _____.

____

____

____

9. Describe how genetics came to be used in animal science.

____

____

10. What is the name for the male parent of an animal?

____

11. _____ What occurs when an individual has a heterozygous genotype and both traits are shown in the phenotype?
    A. Dominant traits
    B. Recessive traits
    C. Codominance
    D. Monohybrids

Key Terms

Carefully study the chapter and then use the word bank to match the term to the correct definition.

A. alleles	E. artificial selection	I. genotype
B. conformation	F. phenotype	J. recessive trait
C. Expected Progeny Difference	G. homozygous	
D. dihybrid cross	H. heterozygous	

1. _____ A probability estimate of the traits shown in offspring, based on two allele pairs
2. _____ Having two different alleles for a single trait
3. _____ Genetic indicators for a trait
4. _____ The process of choosing the most desirable animals for breeding to increase overall animal quality
5. _____ The full summary of an animal's alleles for a specific trait, commonly represented by letters
6. _____ A report of the genetic potential of an animal based on the performance of their offspring
7. _____ These are only shown in crosses where there is no dominant allele present
8. _____ The outward expression of alleles and summary of what you see when you look at an animal
9. _____ Having two of the same alleles for a single trait
10. _____ The overall physical structure of an animal

Critical Thinking

Using the information learned in this chapter, read each question and provide a detailed answer.

1. Describe the differences between genetic selection with crossbreeding and genetic modification.

____

____

____

2. How would livestock be different if genetic selection and crossbreeding were not used in animal agriculture?

____

____

____

Name ______________________ Date ____________ Period ________

Activity: Genetics

Natural Selection Simulation

Required Materials

- Computer with internet access
- Laboratory notebook
- Pen or pencil

Activity Specification

Go to the University of Colorado, Boulder, PhET simulation website. Search for and open the Natural Selection simulation.

Examine how the population of rabbits is impacted by factors such as fur color, food abundance, the presence of a predator, and the temperature.

1. Based on your own observations, describe natural selection.

2. Create a hypothesis that explains how genetic variation plays a role in natural selection.

3. What are two examples of selective pressures that drive changes in the population over time?

4. Analyze how adaptations improve rabbits' chances of survival.

Reflective Questions

Answer the following questions relating to the activity.

1. Analyze the population fluctuations seen in the simulations due to wolf predation. How can this data be used to inform beef cattle ranchers who graze their cattle in an area with a lot of wolf predation?

2. Evaluate the nearest agricultural production commodity to your school. How does natural selection affect its continued production? (Remember, natural selection can take place due to consumer preferences!)

3. Make a prediction regarding the popularity of meat types and how modern consumer preferences will influence their selection.

Name ______________________ Date ____________ Period ________

CHAPTER 7

Reproduction and Parturition

Chapter Review

Carefully study the chapter and answer the following questions.

1. What are the important factors to consider when selecting breeding animals?

2. The tiny hair-like projections that cover the infundibulum are called _____.

3. List three of the most common methods for heat detection.

4. The _____ is where sperm cells mature from spermatids to spermatozoa.

5. Female _____ can store semen in their reproductive tracts for long periods of time to fertilize eggs.

6. The bulbourethral glands are also called _____ _____.

7. The _____ is sometimes called the birth canal.

8. _____ Which of the following are *not* considered seasonally polyestrous?
   A. Mares
   B. Ewes
   C. Does
   D. Cows

9. Explain the vulva's function and what it does to prepare for the birthing process.

10. Identify the correct order of the four stages of the estrous cycle.

A. ______________________ C. ______________________

B. ______________________ D. ______________________

11. Once semen is collected and examined for quality, how is it stored?

____

12. _____ What is the tail of a sperm called?
   A. Acrosome
   B. Flagellum
   C. Motility
   D. Ovum

Key Terms

Carefully study the chapter and then use the word bank to match the term to the correct definition.

A. dystocia
B. flushing
C. motility
D. polyestrous
E. parturition
F. zygote
G. superovulation
H. oogenesis
I. gestation
J. spermatogenesis

1. _____ The cell formed immediately after the egg and sperm nucleus combine
2. _____ The force and path of travel for a sperm cell
3. _____ Difficulty giving birth
4. _____ The period from fertilization of the egg to birth
5. _____ Undergoing estrous regardless of the season
6. _____ The process of stimulating large numbers of follicles and larger-than-normal numbers of eggs produced in ovulation
7. _____ The process where spermatozoa are produced
8. _____ The process where ova are produced in the ovaries of a female animal
9. _____ The process of embryo retrieval through non-surgical means
10. _____ The process of giving birth

Critical Thinking

Using the information learned in this chapter, read the question and provide a detailed answer.

1. Cloning has been viewed as controversial. In your own words, explain the process of cloning. Then, provide pros and cons that could be associated with this process and whether or not you think it is beneficial in livestock production.

____

____

____

____

____

____

____

Name ______________________ Date ______________ Period ____________

Activity: Reproduction and Parturition

Checking Sperm Quality

Required Materials

- Microscope with 1000× magnification
- Microscope slides
- Insemination semen

Activity Specification

Sperm quality is one of the most important attributes of breeding animals via artificial insemination. A farmer must regularly and continuously check the quality of purchased sperm for insemination projects. If a farmer is not receiving healthy sperm, they could be wasting thousands of dollars annually.

In the space provided below, draw a representative sample of sperm as seen under the 1000x magnification feature of your microscope. Also, include a rough percentage estimate of how many of the sperm appear viable.

Sample 1	Sample 2	Sample 3
% Viability =	% Viability =	% Viability =

Goodheart-Willcox Publisher

Reflective Questions

Answer the following questions relating to the activity.

1. Analyze the sperm health/motility in the three samples. Which semen sample is the one you would choose to use and why?

2. Evaluate the perceived ratio of sperm to semen. Which sample appeared to have the highest sperm amount and how did you estimate this?

3. Using the internet, list current market prices for semen from cattle, horses, pigs, sheep, and goats. Why are some more expensive than others?

Name ______________________ Date ____________ Period ________

CHAPTER 8 Food and Fiber Products from Animals

Chapter Review

Carefully study the chapter and answer the following questions.

1. _____ An average American consumes about _____ pounds of meat each year.
   A. 222
   B. 322
   C. 122
   D. 22

2. A term often used when describing meat from the rack is _____.

3. What does the acronym *FMMO* stand for?

4. Describe each of the following USDA egg grades.
   A. USDA Grade AA: ______
   B. USDA Grade A: ______
   C. USDA Grade B: ______

5. What are USDA yield grades and what characteristics are used to grade for yield?

6. Primary products can be broken down into two broad categories: _____ and _____.

7. Today, around _____% of the global population consume meat, and almost _____% consume food from animals.

8. List the five most profitable food products from animals in order.
   A. ______
   B. ______
   C. ______
   D. ______
   E. ______

9. What are beef quality grades and what are the two characteristics used to determine them?

____

____

____

10. What is a similarity between pork and poultry processing that is not relevant to other species?

____

11. _____ are partial breaks on the exterior of the egg that occur prior to laying.

____

12. Milk contains many of the essential nutrients needed for the human diet, including:

_______________ _______________

_______________ _______________

____

13. The USDA has two categories for grading pork: _____ and _____ grades.

____

____

14. _____ What vitamin can only be found in animal-based foods?
   A. Magnesium
   B. B12
   C. Iron
   D. B6

Key Terms

Carefully study the chapter and then use the word bank to match the term to the correct definition.

A. dressing
B. fabrication
C. exsanguination
D. greasy wool
E. homogenization
F. dry aging
G. dressing percentage
H. staple length
I. microns
J. pasteurization

1. _____ The process of removing pathogens from milk
2. _____ Fleeces in their raw form
3. _____ The process of breaking down a side into primal, subprimal, and retail cuts
4. _____ The overall length of fibers in a fleece
5. _____ The process where the processor cuts the jugular vein and bleeds the carcass out
6. _____ The stabilization and mixing of milkfat and other fluid components of milk
7. _____ The process of removing the internal organs, skin, and head
8. _____ The process in which sides or large cuts of beef are left in a refrigerated area and exposed to air for long periods of time
9. _____ Units of measurement for wool fiber diameter
10. _____ The ratio of the carcass weight to the live animal weight

Name ______________________________

Critical Thinking

Using the information learned in this chapter, read the question and provide a detailed answer.

1. Select two species of livestock discussed in this chapter. What similarities and differences are there between the two when looking at processing, cuts of meat, grades, age of harvest, etc.?

Name ______________________ Date ____________ Period ________

Activity: Food and Fiber Products from Animals

Blind Taste Test

Required Materials

- Cooked steak selection (Cornfed: Prime, Choice, Select; Grassfed: Prime, Choice, Select)
- Eating utensils
- Water
- Saltine crackers
- Pen or pencil

Activity Specification

Today you will blindly taste and rank different meat samples. The instructor will cook the following types of meat: Cornfed: Prime, Choice, and Select; and Grassfed: Prime, Choice, and Select. You will receive random samples and must rank them in order of your preference. The following code names have been given to the steaks: red, orange, yellow, green, blue, and violet.

1. **Sample: Red**

   A. **Description:** ______________________

   ____

   ____

   ____

   B. **Final Ranking:** ______________________

2. **Sample: Orange**

   A. **Description:** ______________________

   ____

   ____

   ____

   B. **Final Ranking:** ______________________

3. **Sample: Yellow**

   A. **Description:** ______________________

   ____

   ____

   ____

   B. **Final Ranking:** ______________________

Name ________________________________

4. **Sample: Green**

   A. **Description:** ________________________________

   B. **Final Ranking:** ________________________________

5. **Sample: Blue**

   A. **Description:** ________________________________

   B. **Final Ranking:** ________________________________

6. **Sample: Violet**

   A. **Description:** ________________________________

   B. **Final Ranking:** ________________________________

Reflective Questions

Answer the following questions relating to the activity.

1. Analyze the appearance of the sample meats. How did the appearance of each correlate with its palatability?

2. Evaluate the difference in taste between cornfed and grassfed meat. What might cause this difference?

3. Create an advertisement sign below for a butcher shop and advertise your preferred meat. Emphasize the source of the meat and the flavors that consumers will enjoy.

Name ______ Date ______ Period ______

CHAPTER 9 By-Products from Animals

Chapter Review

Carefully study the chapter and answer the following questions.

1. Animal bones can be heated to a very high temperature and ground into a fine powder that is used for creating beautiful, high-quality _____ called _____ _____.

2. Hides can be used to create the bases for many _____ and _____, while connective tissues found in many hides contain chemicals which can be used as binding agents for _____, _____, and asphalt.

3. Derivatives of _____, _____, and _____ are found in adhesives like glue and paste.

4. Fats are also found in _____, including _____, _____, _____, and _____.

5. Outside of retail meat cuts, the other portions of the animal that can be used include these broad categories:

6. When a beef animal enters the processing plant, more than _____ of the total animal is used in one form or another.

7. What by-product is used to create marshmallows, gummy bears, and Jell-O®?

8. What by-product of wool production is used for lotions, leather conditioner, and as a commercial lubricant for heavy machinery?

____

9. Variety meats come from non-skeletal muscles, including _____, _____, _____ and components of the digestive system.

____

____

10. Explain why animal fats are used for lubrication.

____

Key Terms

Carefully study the chapter and then use the word bank to match the term to the correct definition.

A. tanning
B. leather
C. rendering
D. gelatin
E. pelt
F. lard
G. tallow

1. _____ Cooking fat from beef and pork
2. _____ An odorless, colorless, sterile protein powder that is used in many food products
3. _____ A chemical or mechanical process that preserves the hide of an animal
4. _____ Cooking fat from lamb and mutton animals
5. _____ The process of converting animal by-products into useful goods
6. _____ The resulting product if the hair, wool, or fur is left on the hide and tanned
7. _____ The final product when the hair is removed during the tanning process

Critical Thinking

Using the information learned in this chapter, read each question and provide a detailed answer.

1. Animal by-products are the foundation of many everyday items we use. What products containing animal by-products surprised you?

____

____

____

2. What by-products are used to make the products you listed?

____

____

____

Name ______________________ Date ______________ Period __________

Activity: By-Products from Animals

Making Gummy Candies

Required Materials

- Laboratory notebook
- Laboratory outfit
- Pen or pencil
- Food-grade glassware
- 3.2 ounces of flavored gelatin
- Food dye selection
- ½ ounce unflavored gelatin
- ½ cup corn syrup
- ½ cup water
- Cooking spray
- Small mixed candy molds

Instructions

Today you will make gummy candies! This is possible due to gelatin, which is derived from animal processing. Gelatin is a protein that comes from the connective tissue of animals being harvested. Gelatin assists with thickening semi-liquid and liquid parts of foods.

1. Mix the flavored and unflavored gelatin in a saucepan.
2. Add corn syrup and water to the mixed gelatin saucepan.
3. Mix ingredients over light to medium heat until all gelatin is dissolved.
4. Add desired food dye.
5. Allow ingredients to cool for a few minutes.
6. Spray small mixed candy molds with cooking spray.
7. Carefully pour gelatin mixture into the small mixed candy molds.
8. Place molds into the refrigerator for 20 minutes.
9. Enjoy the candies!

Reflective Questions

Answer the following questions relating to the activity.

1. Analyze the texture of the gummy candies you created and explain how it differs from the raw ingredients used.

2. Evaluate the procedure used to create gummy candies. Why do you think the heating and cooling step changed the texture of the gummy candies to become chewy?

3. What other purposes might gelatin serve?

Name ______________________________ Date ______________ Period __________

CHAPTER 10 Safely Handling Animals

Chapter Review

Carefully study the chapter and answer the following questions.

1. Explain how flight zones, pressure zones, and point of balance influence livestock behavior and how people work safely with animals.

____

____

2. Which process is rarely a good option to use for control of a cat if they are being calm or do not require additional restraint?

____

3. Crossing behind a horse is best done either very close to the horse with a hand on its _____, or _____ _____ behind the horse.

____

____

____

4. Describe ways you can show dogs you are not a threat to them.

____

____

____

____

____

5. _____ are handled less frequently than other livestock species.

____

6. _____ _____ are considered extremely docile and easy to handle compared to other species due to genetics and increased human handling.

____

____

7. Livestock have less ability to perceive _____ or _____ than humans do.

____

____

8. _____ are more likely to be curious and approach people than _____, and are also more likely to jump when they feel threatened.

____

____

9. _____ are the most dangerous of livestock species.

10. List the main safe handling principles outlined in the chapter.

Key Terms

Carefully study the chapter and then use the word bank to match the term to the correct definition.

A. imprinting	D. muzzle	G. cat bag
B. scruffing	E. head gate	H. stocks
C. squeeze chute	F. control pole	I. snare

1. _____ A rope loop placed around a pig's upper jaw and then looped over a fence
2. _____ A form of restraint that slides behind the head of a cow to prevent it from backing up
3. _____ The process of picking a cat up by the loose skin at the back of its neck
4. _____ Can be slipped over a dog's neck to allow for control at a distance
5. _____ A piece of equipment which is fastened to a small animal's head to prevent it from biting
6. _____ A process in which newborn horses are exposed to human handling
7. _____ A specially designed bag which allows handlers to expose only the area that needs to be accessed while keeping the rest of the cat secure
8. _____ A standing chute for horses
9. _____ A form of restraint that can calm cattle as the entire body feels gentle pressure while keeping the animal still

Critical Thinking

Using the information learned in this chapter, read the question and provide a detailed answer.

1. Livestock handling is incredibly important for the safety and well-being of the humans and animals involved. Why? What long-term effects are the result of poor handling practices?

Name ______________________ Date ____________ Period ________

Activity: Safely Handling Animals

Handling Cattle

Required Materials

- Lab notebook
- Closed-toed shoes
- Long pants

Activity Specification

Today you will explore handling cattle. You will do this by recognizing the flight zones, blind spots, and handling positions of cattle. You will work in groups of three to navigate a bovine from one specific location to another, through a course designed by the instructor. At all times, you must be calm, safe, and wear appropriate clothing.

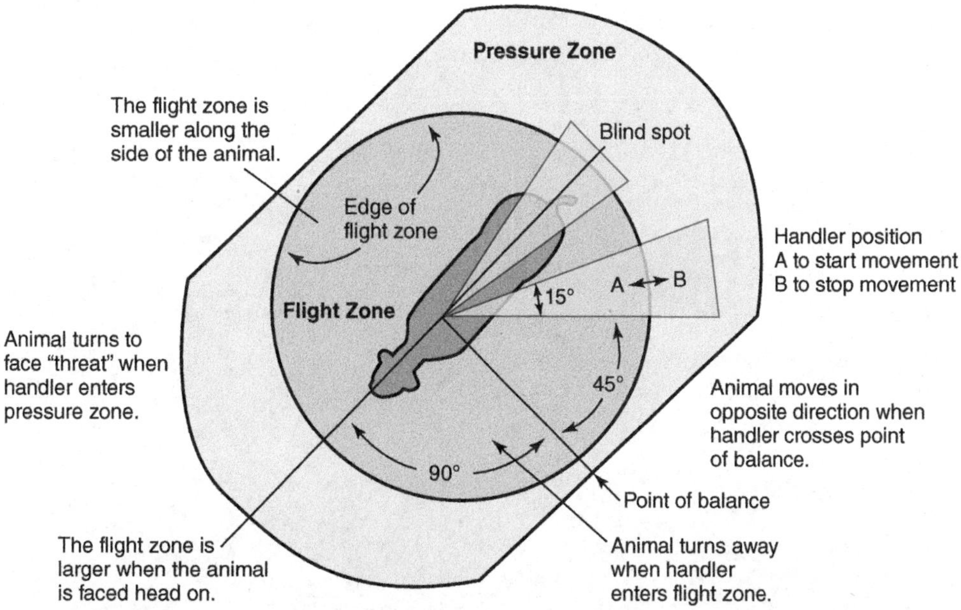

Goodheart-Willcox Publisher

Reflective Questions

Answer the following questions relating to the activity.

1. Analyze the eye and ear movements of cattle as your classmates walk around them. How do their eyes and ears react to the presence of people?

2. Evaluate forelimb and hindlimb mechanics as the cattle adjust their positioning when people are present. How can walking patterns help you understand the direction a bovine wants to move in?

3. Create a poster below that educates new students about the proper manner to direct cattle movement.

Name ______________________ Date ____________ Period ________

CHAPTER 11 Animal Welfare

Chapter Review

Carefully study the chapter and answer the following questions.

1. The Animal Welfare Act (AWA) was originally passed in _____.
2. What is the purpose of the Five Freedoms and how did they come to be?
3. Name the Five Freedoms associated with animal welfare.
4. How is the AWA enforced and how are deficiencies handled if found?
5. Which of the Five Freedoms is addressed through management practices like deworming?
6. Which of the Five Freedoms is most applicable to safe animal handling practices?
7. What regulations are outlined for each category of animal in the Animal Welfare Act?

8. _____ Which of the following animals are not covered under the AWA?
   A. Fish
   B. Farm animals used for food or fiber
   C. Amphibians and reptiles
   D. None of these are covered by the AWA

9. The Humane Methods Slaughter Act was originally passed in _____.

____

10. Which of the Five Freedoms is the simplest concept related to animal welfare?

____

____

Key Terms

Carefully study the chapter and then use the word bank to match the term to the correct definition.

A. animal husbandry	C. Five Freedoms	E. animal rights
B. animal welfare	D. Humane Methods Slaughter Act	F. Animal Welfare Act

1. _____ The main federal legislation guiding animal welfare for domestic animals in the United States
2. _____ The guiding principles of animal welfare
3. _____ A philosophical view that animals have rights similar to humans
4. _____ Legislation that regulates the welfare of animals produced for human consumption
5. _____ The care of animals that are raised for meat, clothing, or other products for human consumption
6. _____ A philosophical view which advocates for the responsible care of animals while allowing animals to be used for traditional purposes

Critical Thinking

Using the information learned in this chapter, read the question and provide a detailed answer.

1. Imagine you encounter a peaceful animal rights protest that is lobbying against animal agriculture. They are concerned with how livestock are treated, their housing conditions, and slaughter treatment. What would your response be to these concerns after reading this chapter?

____

____

____

____

____

____

____

____

____

____

____

____

Name ______________________ Date ____________ Period ________

Activity: Animal Welfare

Depicting Animal Rights and Animal Welfare

Required Materials

- Computer with internet access
- Gmail account

Activity Specification

Today you will create an infographic that clarifies and distinguishes between animal rights and animal welfare.

1. Go to the Piktochart website.
2. Log in with a Gmail account.
3. Click on Infographics.
4. Select a free template or develop your own.

You must articulate details that distinguish animal rights and animal welfare with simple statements and images. You will print out this infographic in an 8.5" × 11" document which will be posted in your classroom. You will deliver a 30-second elevator pitch to explain the contents of the poster to your classmates.

Reflective Questions

Answer the following questions relating to the activity.

1. Analyze distinguishing features between animal rights and animal welfare. Why do you believe it is important for the public to be educated on the distinction between the two?

2. Evaluate your peers' posters and verbal presentations. Write down two phrases from these presentations that stood out to you as being most effective for educating others about animal rights and animal welfare.

3. Create a four- to five-step protocol for calmly addressing concerns of an animal rights protester at a local livestock fair.

Name ______________________ Date ____________ Period ________

CHAPTER 12 Animal Science and the Scientific Method

Chapter Review

Carefully study the chapter and answer the following questions.

1. _____ occurs when scientists study the living and physical world and work toward well-established explanations that describe the phenomena they observe.

____

2. A(n) _____ is generally a 250-word or shorter overview of an experiment.

____

3. _____ is critical to continued scientific development because it demonstrates that scientific inquiry is a continuous process, supported by many, and attempts to address the limitations of scientific advancements.

____

4. More than 50 million years ago, horses were the size of small _____.

____

5. Molecules combine to form _____.

____

6. Describe the difference between basic research and applied research.

____

____

7. Describe the difference between precision and accuracy.

____

____

8. List the steps of the Scientific Method from start to finish.

A. ____

B. ____

C. ____

D. ____

E. ____

F. ____

G. ____

9. Animal science is the area of study connecting animal _____ and _____ to application.

____

____

10. The _____ is where the authors explain why their experiment turned out as it did and how their results add to the growing body of a theory or a specific discipline.

____

11. How does vaccination differ from variolation?

____

12. A _____ is well established and supported by considerable scientific evidence but cannot be proven true and thus is not immune to change.

____

Key Terms

Carefully study the chapter and then use the word bank to match the term to the correct definition.

A. accuracy	E. independent variable	I. peer-review process
B. biomes	F. qualitative data	J. quantitative data
C. statistical evaluation	G. sample size	
D. dependent variable	H. ecosystem	

1. _____ Employs mathematical calculations that infer results for populations from representative sample populations
2. _____ Number of participants involved in an experiment
3. _____ The collection of all the biotic and abiotic attributes of all organisms located in a specific area
4. _____ Refers to the degree of conformity to a true value
5. _____ Observations that are categorical or descriptive in nature
6. _____ The process through which experts in science review research submissions and question the experimental design and interpretation of findings
7. _____ The variable that a scientist manipulates in an experiment
8. _____ Groups of similar ecosystems
9. _____ Data that is expressed numerically
10. _____ The variable that will be evaluated as evidence

Critical Thinking

Using the information learned in this chapter, read the question and provide a detailed answer.

1. Think about something that you are curious about and that you would be interested in researching within animal science. Then, follow the steps of scientific inquiry as if you were going to carry out an experiment. What would you investigate? What methods would you use? How would you start?

____

____

____

____

____

____

____

____

Name ______________________ Date ____________ Period ________

Activity: Animal Science and the Scientific Method

Practicing Data Collection

Required Materials

- Lab notebook
- Pen or pencil
- 12-inch ruler
- Meter stick
- Large livestock scale
- Small livestock scale
- Digital thermometer
- Proper laboratory or farm outfit

Activity Specification

Successful experiments rely on proper data collection methods. Today, you will measure the length, weight, volume, and temperature of various items and animals in the school farm or laboratory.

Measure the length, width, height, and volume of the following pens:

	Swine/Lamb/Goat Pen	Rabbit Pen	Cattle Pen
Length (meters)			
Width (meters)			
Height (meters)			
Volume (meters^3)			

Your instructor will place various graduated cylinders around the farm with animal urine for analysis. Write down the amount of urine in each graduated cylinder and its temperature.

	Graduated Cylinder #1	Graduated Cylinder #2	Graduated Cylinder #3	Graduated Cylinder #4
Volume (mL)				
Temperature (°C)				

Around the farm, find ANY three items/objects (ex. goat feces, walking stick, etc.) that are appropriate to measure (length) with the following three measurement units:

	Item #1 Measure in Meters	Item #2 Measure in Centimeters	Item #3 Measure in Millimeters
Item Name			
Item Length			

Reflective Questions

Answer the following questions relating to the activity.

1. Analyze the behavior of livestock within their pens. How might pen size or design relate to perceived animal stress?

2. Evaluate the common walkways in the farm. List two to three safety risks that can pose danger for students collecting data for research.

3. Identify an area of the farm where you think changes can improve animal productivity. Create a hypothesis explaining how and why changing this could improve productivity.

Name ______________________ Date ____________ Period ________

CHAPTER 13 Beef Industry Overview

Chapter Review

Carefully study the chapter and answer the following questions.

1. _____ Worldwide, beef is the _____ most widely consumed meat.
   A. first
   B. second
   C. third
   D. fourth

2. Cattle are large ruminant animals with _____ hooves that are used to produce meat, milk, or work.

   ____

3. _____ _____ cattle were domesticated in Europe.

   ____

   ____

4. There are approximately _____ head of cattle in the United States and the average herd size is _____ head per producer.

   ____

   ____

5. Less expensive retail cuts of beef are cut from which primal cuts?

   ____

   ____

   ____

   ____

6. _____ _____ is credited with first bringing cattle to North America in _____.

   ____

   ____

   ____

7. _____ is the scientific name for beef cattle.

   ____

8. _____ _____ cattle were originally bred in India.

   ____

   ____

9. There are more than _____ recognized cattle breeds raised commercially in the United States.

   ____

10. Early cattle were used primarily as _____ _____.

11. The development of the _____ revolutionized the beef industry in the 1880s.

12. _____ _____ cattle were descendants of the aurochs of Asia, Europe, and North Africa.

13. The beef cattle industry generates _____ in cash receipts annually by selling cattle and calves.

Key Terms

Carefully study the chapter and then use the word bank to match the term to the correct definition.

A. loin
B. heifer
C. round
D. calving
E. chuck
F. bull
G. calf
H. cow
I. rib
J. steer

1. _____ The primal cut of meat from the hindquarters of beef cattle
2. _____ Young cattle (male or female), typically younger than six months of age
3. _____ Intact male beef animal of reproductive age
4. _____ Act of giving birth to cattle
5. _____ The primal cut from the section between the ribs and the round, above the flank
6. _____ Castrated male beef animal
7. _____ The primal cut from the shoulder and neck of beef cattle
8. _____ Mature female beef animal that has calved
9. _____ The tender primal cut made from the center section of a rib
10. _____ Young female beef animal that has not calved

Critical Thinking

Using the information learned in this chapter, read the question and provide a detailed answer.

1. What do you think is the most influential turning point that led to today's vibrant beef industry? Explain your choice using examples from the text.

Name ______________________________ Date ______________ Period __________

Activity: Beef Industry Overview

Describing Sectors of the Beef Industry

Required Materials

- Computer with internet access
- Gmail account

Activity Specification

Today you will create an infographic that defines and distinguishes the four main sectors of the beef cattle industry: seedstock operations, cow-calf operations, stocker operations, and feedlot operations.

1. Go to the Picktochart website.
2. Log in with a Gmail account.
3. Click on Infographics.
4. Select a free template or develop your own.

You must articulate details that distinguish one type of operation from another with simple statements and images. You will print out this infographic in an 8.5" × 11" document which will be posted in the classroom.

Reflective Questions

Answer the following questions relating to the activity.

1. Analyze the purpose of each segment of the beef industry. What recommendations do you have for facilitating the physical and financial transfer of cattle products between segments?

   ____

   ____

   ____

2. Research ingredients typically found in feedlot diet rations. Evaluate these contents and identify why they are included.

   ____

   ____

   ____

3. Create a blueprint for a feedlot facility. Clearly identify the cattle beds, feed troughs, water troughs, shaded areas, fans, paths for tractors, and any other aspects requested by your instructor.

Name ______________________ Date ____________ Period ________

CHAPTER 14 Beef Cattle Production Systems

Chapter Review

Carefully study the chapter and answer the following questions.

1. The gestation period for cattle is between _____ and _____ days and varies a bit depending on the _____ of cattle.

2. _____ percent of beef cattle consumed in the United States is finished on a feedlot.

3. Stocker operations can be found across the United States, but have shifted more toward the _____ region.

4. Cattle that are easy to handle are:

5. The average size of a cow-calf operation in the United States is _____ head.

6. How does herd behavior differ from individual behavior?

7. List the performance data that breed associations document.

8. Heifers are typically bred when they are _____ to _____ months old so they calve for the first time at approximately _____ years of age.

9. Describe a disadvantage of using video auctions to sell cattle.

____

____

10. Feedlot operations sort cattle into groups based on their _____ and _____.

____

____

11. Market cattle are raised to a weight of _____ to _____ pounds before they are harvested.

____

____

12. _____ is the number one state in cow-calf production.

____

13. Most market cattle are harvested between _____ and _____ months of age.

____

____

14. How do private treaty sales help to maximize profits?

____

____

Key Terms

Carefully study the chapter and then use the word bank to match the term to the correct definition.

A. seedstock	D. weaned	G. confinement
B. backgrounding operation	E. purebred	H. herd
C. feedlot	F. cow-calf operation	

1. _____ Group of cattle that stays together
2. _____ An operation in which cows are bred each year to produce calves for market, with some calves retained as replacement animals
3. _____ Occurs when a six-month-old calf is separated into a different pen or pasture from its mother
4. _____ An operation in which large numbers of cattle are sorted into groups based on their size and genetics and fed to reach market weight
5. _____ An animal whose parents belong to the same breed and whose ideal characteristics are achieved through the process of selective breeding
6. _____ An operation in which weaned calves of both sexes are sent to graze and gain weight prior to being sent to a feedlot
7. _____ Being managed in a controlled area and not allowed to graze
8. _____ Bulls, heifers, and cows that are used for reproduction in registered breeding herds as well as commercial cattle operations

Name ____

Critical Thinking

Using the information learned in this chapter, read the question and provide a detailed answer.

1. What are some things that a beef cattle producer would need to do to prepare for a private treaty sale? Why are these things important, and how could they affect the producer's profit?

____

____

____

____

____

____

Name ______________________ Date ______________ Period __________

Activity: Beef Cattle Production Systems

Evaluating Rangeland

Required Materials

- Lab notebook
- Pen or pencil
- 5 resealable bags
- 1ft. × 1ft. soil quadrat
- Access to rangeland area

Activity Specification

How do you evaluate whether an area is appropriate to raise beef cattle? What do people raising cattle need to consider when looking at an area? In an open field, toss a soil quadrat four times in different directions. The position you stand at is called the Point of Origin.

Samples	% of Vegetation*	% of Exposed Soil or Rocks*	Animals or Signs of Animals	Water/Moisture Levels	Weather
Sample #1 (Point of Origin)					
Sample #2					
Sample #3					
Sample #4					
Sample #5					

*Collect a small sample to bring back to the classroom for analysis. Place in a resealable bag.

Name ____________________

Reflective Questions

Answer the following questions relating to the activity.

1. Analyze the vegetation and soil contents in your samples. Describe the sizes and textures of at least three of these components.

2. Evaluate all samples collected by your class. What common themes are present?

3. What is necessary to raise beef cattle? Would your area be well-suited to raising cattle? Why?

4. What makes for ideal rangeland?

Notes

Name ________________________________ Date ______________ Period __________

CHAPTER 15 Beef Cattle Management and Care

Chapter Review

Carefully study the chapter and answer the following questions.

1. What illness is caused by *Theileria* protozoa?

____

2. What viral cattle illness can be treated by vaccination and includes symptoms like fever, lethargy, diarrhea, lack of coordination, and failure to nurse calves?

____

3. Body condition score ranges from _____ to _____.

____

____

4. List the three classes of dewormers.

______________________________ ______________________________

____

5. What illness is caused by overconsumption of large amounts of carbohydrate-rich, highly fermentable feeds at one time?

____

6. What bacteria causes scours in calves?

____

7. _____ BCS scores indicate nutritional issues in the herd.

____

8. There are _____ minerals required to meet beef cattle's nutritional needs.

____

9. What parasite is carried on snails and ingested by grazing cattle?

____

10. List the common forms of killed vaccines and toxoids.

______________________________ ______________________________

______________________________ ______________________________

11. When lactating, a beef cow can consume approximately _____ pounds of _____ each day.

____

____

12. List the three main types of identification used for beef cattle.

________________________ ________________________

____

13. What illness is spread by infected animals passing the protozoa eggs in their fecal matter, and then picked up by other animals that ingest contaminated food or water?

____

14. What illness causes cattle urine to turn dark red?

____

15. What respiratory disease is carried by biting midges (flies) and spreads from animal to animal as infected midges bite them?

____

Key Terms

Carefully study the chapter and then use the word bank to match the term to the correct definition.

A. forager
B. ruminant
C. killed vaccine
D. breech
E. structural correctness
F. body condition score (BCS)
G. modified live vaccine
H. blackleg
I. bovine spongiform encephalopathy (BSE)

1. _____ Birthing position in which the calf is turned completely backwards with its back feet presenting first
2. _____ Animal that takes in forage and ferments it in its multi-part stomach before regurgitating it to chew and digest
3. _____ Estimates the body composition of breeding cattle and accounts for the fat reserves the cattle may draw from
4. _____ An infectious, non-contagious disease caused by ingesting *Clostridium chauvoei*
5. _____ How cattle are built and evaluated, in terms of their skeleton
6. _____ These, along with toxoids, have organisms or parts of organisms that do not replicate once injected or administered into the animals
7. _____ Animal that roams across grasslands in search of food
8. _____ A prion disease that is typically passed through contaminated feed, containing the remnants (meat and bone meal) of infected animals
9. _____ These contain small amounts of a live virus or bacteria

Critical Thinking

Using the information learned in this chapter, read the question and provide a detailed answer.

1. Select an illness mentioned in this chapter that interests you. What are the symptoms you would need to look for? How can you prevent this illness as part of a herd health program?

____

____

____

____

____

Name ______________________ Date ____________ Period ________

Activity: Beef Cattle Management and Care

Freeze Branding

Required Materials

- Liquid nitrogen storage tank
- Branding irons
- Gloves
- Leather hides
- Goggles
- Lab coat
- Lab notebook

Activity Specification

Freeze branding, also known as cold branding, is a method of branding cattle without using heat. This method is less painful to cattle than hot branding and can also provide a clearer brand. Today you will be able to safely practice this procedure on leather hides.

1. Put on all personal protective equipment (gloves, goggles, lab coat).
2. Open the liquid nitrogen storage tank and insert the branding iron in cold liquid nitrogen (colder than -300°F) for 7 to 10 minutes with a closed lid.
3. Carefully remove the branding iron and firmly press it onto the leather hide for 25 seconds.

Reflective Questions

Answer the following questions relating to the activity.

1. Analyze the leather hides after they have been branded and the temperature warms up. Describe the texture and appearance of the brand.

____

____

____

2. Evaluate the clarity of the brand. What do you think can be done to improve cold branding?

____

____

____

3. Create a personalized brand below.

4. Describe why you chose to design the brand in this manner.

Name ______________________ Date ____________ Period ________

CHAPTER 16 Beef Cattle Selection and Evaluation

Chapter Review

Carefully study the chapter and answer the following questions.

1. _____ _____ are used to produce market cattle or future breeding cattle.

2. What characteristics make an ideal cow or heifer?

3. _____ _____ are raised for the purpose of producing meat for human consumption.

4. Where do most structural issues occur in cattle?

5. List the main considerations a producer will take into account in selecting market cattle from most important to least important.

   A. ____________ D. ____________

   B. ____________ E. ____________

   C. ____________ F. ____________

6. _____ is a producer's careful evaluation and addition of animals to a herd that bring desired characteristics.

7. What must be done for performance data to be a useful and effective tool?

8. List the main considerations producers should take into account when selecting breeding cattle, from most important to least important.

   A. ____________ D. ____________

   B. ____________ E. ____________

   C. ____________ F. ____________

9. To _____ is to make a judgement about the _____, _____, or _____ of something.

10. What characteristics make an ideal bull?

Key Terms

Carefully study the chapter and then use the word bank to match the term to the correct definition.

A. splayfooted
B. cutability
C. post-legged
D. sickle-hocked
E. calf-kneed
F. bowlegged
G. buck-kneed
H. balance
I. degree of finish
J. frame score

1. _____ Hyperextended knees
2. _____ Proportional in all areas and symmetrical
3. _____ The amount of fat covering found on the beef animal
4. _____ Hind leg that has too much curvature in relation to how the leg is set
5. _____ Knees set too far forward
6. _____ A measurement of the percentage of boneless retail cuts from the round, loin, rib, and chuck; related to yield grade
7. _____ Hind legs that are too straight
8. _____ A measurement of an overall size, which is determined by measuring cattle from the ground to their hooks and factoring in the animal's age
9. _____ Toes that point outward rather than forward
10. _____ Legs that flare out at the hock and then narrow down to the pastern and hoof

Critical Thinking

Using the information learned in this chapter, read the question and provide a detailed answer.

1. Why is structural correctness vital for livestock? How can producers ensure structural soundness is maintained throughout their herd over time?

Name ______________________ Date ____________ Period ________

Activity: Beef Cattle Selection and Evaluation

Showing Beef Cattle

Required Materials

- Laboratory notebook
- Pen or pencil
- Farm or laboratory outfit
- Halter
- Show stick

Activity Specification

Beef cattle showmanship is an activity performed by members of agriculture organizations such as FFA and 4-H. They often show their animals at public livestock shows. Today, your instructor will demonstrate the basics of showing beef cattle, or have you watch a livestock show.

NOTES:

Reflective Questions

Answer the following questions relating to the activity.

1. Analyze the behavior of cattle as they are being led across the arena. What are three factors that might distract their performance?

2. Evaluate the body posture of the student showing the animal. What are three attributes of their posture that stand out to you?

3. Create a list of five tips that new handlers can employ to show their cattle effectively.

Notes

Name ______________________ Date ____________ Period ________

CHAPTER 17 Classification and Breeds of Beef Cattle

Chapter Review

Carefully study the chapter and answer the following questions.

1. There are more than _____ breeds of cattle recognized worldwide.

2. This German breed uses the term "balancer" to describe animals that have 25%–75% of the breed in their pedigree.

3. This breed was developed in Texas and produces hardy, disease-resistant cattle that easily adapt to many different production scenarios.

4. What made continental breeds increase in popularity in the 1960s and 1970s?

5. _____ Which of the following is *not* considered a *Bos indicus* breed?
   A. Devon
   B. Brahman
   C. Braford
   D. Beefmaster

6. This breed can be traced back to 1912 at a USDA Experiment Station in Louisiana.

7. This breed was originally known as Durham.

8. _____ is the name for all species of cattle.

9. What makes *Bos indicus* breeds appealing to producers in the Southeastern United States?

10. This breed dates to the Middle Ages and is one of the oldest and most widely distributed breeds of cattle in the world.

11. Originating from Ireland, this breed is believed to derive from Kerry and Devon cattle.

12. What are the five most popular continental breeds in the United States today?

13. This breed originated in Scotland and was imported to the United States by George Grant in 1873.

____

14. This breed has black pigmented skin and is also referred to as zebu.

____

15. _____ Which of the following breeds is *not* considered British?
    A. Angus
    B. Shorthorn
    C. Simbrah
    D. Hereford

Key Terms

Carefully study the chapter and then use the word bank to match the term to the correct definition.

A. terminal characteristics
B. breed association
C. *Bos taurus*
D. *Bos indicus*
E. breed
F. polled
G. maternal characteristics

1. _____ Characteristics that lead to success in breeding and raising cattle, such as calving ease, yearling weight, and milk production
2. _____ Breeds of cattle that originated in Europe
3. _____ A specific group of animals that share a similar appearance
4. _____ Organization that manages the registration of purebred cattle
5. _____ Breeds of cattle from warm climates in Asia and Africa
6. _____ Characteristics that lead to optimal beef production, such as muscling, degree of fat, carcass yield, and growth
7. _____ Hornless

Critical Thinking

Using the information learned in this chapter, read the question and provide a detailed answer.

1. Explain the difference between terminal and maternal characteristics, then give three examples of each.

____

____

____

____

____

____

Name ______________________ Date ____________ Period ________

Activity: Classification and Breeds of Beef Cattle

Beef Cattle Breed Comparison

Required Materials

- Laboratory notebook
- Pen or pencil
- Computer with internet access

Activity Specification

Many reasons exist for the selective breeding of cattle including, but not limited to, enhancing productivity, disease resistance, carcass quality, environmental adaptability, consumer demand, and for the conservation and improvement of genetic diversity.

1. In your own words, describe heterosis (hybrid vigor).

____

____

2. Go to the National Association of Animal Breeders website. List four breeds of beef cattle and at least one important attribute of each.

______________________ ______________________

______________________ ______________________

3. Paste images of the four beef breeds discussed above.

Breed #1:

Breed #3:

Breed #2:

Breed #4:

Reflective Questions

Answer the following questions relating to the activity.

1. Analyze the different structure/shape of the breeds you have chosen. What similarities and differences can you identify?

2. Evaluate breeds and their common purpose. Pick one of your selected breeds and explain how its genetics are associated with its purpose.

3. Select your favorite beef breed and create an infographic with an image and at least five different facts about this breed below.

Name ______________________ Date ____________ Period ________

CHAPTER 18 Swine Industry Overview

Chapter Review

Carefully study the chapter and answer the following questions.

1. The United States exports $_____ of pork, accounting for _____% of worldwide pork exports.

2. Swine are considered ready to harvest at approximately _____ pounds.

3. Which primal cut is often used to make sausage or other ground pork products?

4. _____ cuts are cut down into the _____ cuts that consumers buy at the grocery store.

5. Cuts from this primal are often smoked and are fattier than other cuts.

6. _____ has the highest per capita pork consumption in the world, followed by _____ and the _____ _____.

7. What is the scientific name for swine?

8. According to the 2017 Census of Agriculture, the US swine industry generated $_____ in sales, which accounts for _____% of total annual agricultural sales in the United States.

9. The _____ is an organization that advocates for and supports pork producers and represents _____ state associations.

10. Swine have a dressing percentage of _____%.

11. _____ is the largest importer of US pork.

12. The chops, roasts, and ground meat commonly used as a source of protein in our diets are cut mainly from the _____, _____, and _____.

13. List the top five states, in order, in US swine production.

A. ______ D. ______

B. ______ E. ______

C. ______

Key Terms

Carefully study the chapter and then use the word bank to match the term to the correct definition.

A. swine
B. ham
C. loin
D. market hog
E. piglet
F. sow
G. flitch
H. gilt
I. barrow
J. boar

1. _____ Primal cut made along the top of the animal, between the shoulder and the leg
2. _____ Mature female hog that has reproduced
3. _____ Intact male hog of any age
4. _____ A full side of cured or uncured bacon
5. _____ An animal intended for slaughter at 280 pounds
6. _____ Castrated male hog
7. _____ The primal cut found at the back of the hog's body, encompassing the fleshy top part of the rear leg to the top of the animal
8. _____ Name for hogs at various stages of development
9. _____ Young female hog that has not reproduced
10. _____ Young swine (male or female)

Critical Thinking

Using the information learned in this chapter, read the question and provide a detailed answer.

1. What are some of the major advantages of pork production compared to beef production? How do these factors contribute to its greater global popularity than beef?

Name ______________________ Date ____________ Period ________

Activity: Swine Industry Overview

Handling Swine

Required Materials

- Laboratory notebook
- Pen or pencil
- Farm or laboratory outfit
- Pig board

Activity Specification

Today your instructor will show you how to handle swine using pig boards. Your instructor has swine in a pen and will establish a location where they expect you to walk the swine. You will work in pairs to move your animal from point A to point B. In the image below, copy down the course that your instructor wants you and your partner to guide your swine through.

Goodheart-Willcox Publisher

Reflective Questions

Answer the following questions relating to the activity.

1. Analyze the head of a pig, with particular attention to its eyes and snout. How do pigs' eyes and snouts contribute to their movement?

2. Evaluate the ease or difficulty associated with moving swine. What techniques worked the best? What did not work well?

3. What are your top three strategies to successfully and safely guide swine?

Name ____________________ Date __________ Period ______

CHAPTER 19 Swine Production Systems

Chapter Review

Carefully study the chapter and answer the following questions.

1. This crossbreeding system produces a maximum number of uniform animals for market.

____

2. Boars typically reach breeding age at _____ to _____ months.

____

____

3. List the three common types of crossbreeding systems for hogs.

_______________ _______________

____

4. The gestation period for female hogs is approximately _____ days.

____

5. Why is shade particularly essential for pigs raised in an outdoor system?

____

6. _____ operations are the most common type of commercial operation.

____

7. Which crossbreeding system rotates the breed sires in each generation?

____

8. _____ is the term used for the process of pigs giving birth.

____

9. List the three types of confinement operations.

_______________ _______________

____

10. The litter size for hogs typically falls between _____ and _____ piglets.

____

____

11. Farrow-to-wean operations are also called _____.

____

12. What is the main disadvantage of the rotaterminal crossbreeding system?

____

Key Terms

Carefully study the chapter and then use the word bank to match the term to the correct definition.

A. effluent
B. confinement operations
C. agonistic
D. all in-all out
E. rotaterminal crossbreeding
F. concentrated animal feeding operation
G. farrow-to-finish
H. farrow-to-wean
I. finishing operation

1. _____ Combative
2. _____ Waste converted to liquid form and pumped into secured lagoons outside the facility where bacteria break down the feces and urine to be used as fertilizer
3. _____ Large-scale or commercial swine production where the pigs are bred, born, and raised in large indoor buildings
4. _____ System in which crossbred females are bred to boars of different breeds in a rotating pattern, with some females being kept as replacement gilts while the remainder go to market
5. _____ Operation in which sows and gilts farrow and nurse their litters, and then the feeder pigs are sold to a finishing operation
6. _____ Intensive animal feeding operation where more than 1000 animals are located for more than 45 days per year
7. _____ The practice of moving groups of swine of the same size, age, and gender through the production process together
8. _____ Operation in which sows or gilts farrow year-round
9. _____ Operation in which weaned piglets gain weight before being sent to market

Critical Thinking

Using the information learned in this chapter, read each question and provide a detailed answer.

1. Describe the reasons why swine operations confine the animals.

____

____

____

____

2. Describe the reasons some swine operations use farrowing crates.

____

____

____

____

Name ______________________ Date ______________ Period __________

Activity: Swine Production Systems

Checking Swine Temperature

Required Materials

- Laboratory notebook
- Pen or pencil
- Pigs
- Farm or laboratory outfit with gloves
- Thermometer
- Lubricant

Activity Specification

Today you will demonstrate your ability to temperature-check swine. Using the pig-boarding skills from the previous lesson, in a group of three, two students will secure a pig in a corner while a third takes its temperature using a rectal thermometer.

1. Two students will pig-board a pig calmly to a corner in a pen.
2. The third student will apply lubricant to a digital thermometer tip.
3. Place the thermometer tip in the rectum of the animal.
4. Record the temperature.
5. Allow the pig to return to its companions.
6. Repeat the process until three measurements are obtained.

Temperature #1 [°F]	Temperature #2 [°F]	Temperature #3 [°F]	Average Swine Temperature [°F]

Reflective Questions

Answer the following questions relating to the activity.

1. Analyze the temperatures of the swine that were sampled. How do they compare to the ideal temperature of swine?

2. Evaluate the average swine temperatures collected in your class and explain reasons for the deviations of temperature obtained compared to the ideal temperature of swine.

3. Create a testable hypothesis that describes why the appropriate internal temperature of swine is several degrees higher than for humans.

Name ______________________ Date ____________ Period ________

CHAPTER 20 Swine Management and Care

Chapter Review

Carefully study the chapter and answer the following questions.

1. List the preferred forms of identification used in swine production.

______________________ ______________________

____

2. _____ and _____ are the best ways to keep healthy swine herds.

____

____

3. This painful skin condition has unknown causes and is exacerbated by stress and sun exposure.

____

4. Swine will lay down most of the day and can sleep as many as _____ hours per day.

____

5. A quality herd management plan will include the following:

______________________ ______________________

______________________ ______________________

______________________ ______________________

____

6. This illness presents symptoms such as muscle tremors, rapid respiration, facial twitching, and sudden death.

____

7. Producers employing artificial insemination should complete the process _____ to _____ hours after the detection of estrus and again _____ hours later.

____

____

____

8. Swine internal anatomy is the closest to _____ of any food production animals.

____

9. Processing piglets includes completing the following tasks:

______________________ ______________________

______________________ ______________________

______________________ ______________________

10. Body condition scores for swine range from _____ to _____.

11. Atrophic rhinitis is caused by two primary organisms, _____ or _____.

12. Primarily observed in young pigs, this disease is caused by *Staphyloccocus hyicus* bacteria that form colonies on broken skin and cause inflammation, blistering, and fluid seepage.

13. Swine are especially susceptible to _____ infections.

14. A boar should not _____ more than seven times per week.

15. Penicillin is a typical treatment for this illness that can cause diamond-shaped lesions.

Key Terms

Carefully study the chapter and then use the word bank to match the term to the correct definition.

A. anthelmintic
B. flushing
C. standing heat
D. cloven-hoofed
E. vaccination schedule
F. weaning

1. _____ When the female is sexually receptive to the male and will stand to be mounted
2. _____ Also called dewormers, these are used to control parasites
3. _____ Having hooves that are split into two toes
4. _____ A series of vaccinations intended to maintain herd health
5. _____ The practice of over-feeding gilts to enhance the chances of conception
6. _____ The process of separating young animals from their mothers

Critical Thinking

Using the information learned in this chapter, read each question and provide a detailed answer.

1. Select an illness covered in this chapter that interests you and describe how it develops in pigs and the symptoms to watch for.

2. What herd health protocols would be especially important to prevent the illness you selected?

Name ______________________ Date ______________ Period __________

Activity: Swine Management and Care

Practicing Swine Insemination

Required Materials

- Laboratory notebook
- Pen or pencil
- Farm or laboratory outfit with gloves
- Narrow desk
- Bucket of water
- Semen bag with water
- Swine AI rod
- Lubricant
- Breeding saddle

Activity Specification

Today you will learn how to secure and assist with the insemination of a pig. Your instructor will demonstrate the actual insemination, but you will practice the set-up and assistance portion with a faux set-up.

1. Apply lubricant to the swine AI rod.
2. Attach the semen bag to the AI rod.
3. Securely attach the breeding saddle to a narrow desk (the narrow desk simulates a pig who will be inseminated).
4. Attach the AI rod and semen bag combination to the breeding saddle.
5. Firmly secure the breeding saddle to the narrow desk, and lightly pump water through the rod.
6. Collect the water in the bucket.

Reflective Questions

Answer the following questions relating to the activity.

1. Analyze the tip of the swine AI rod. Why is it this shape and do you expect this to influence the degree of success in insemination?

   ____

   ____

   ____

2. Evaluate the insemination process. What are two tips that you can offer a farmer who needs to inseminate their pigs quickly and cleanly?

   ____

   ____

   ____

3. Create an image below that shows an AI rod correctly saddled to a receiving gilt or sow.

Name ____________________ Date __________ Period ______

CHAPTER 21 Swine Selection and Evaluation

Chapter Review

Carefully study the chapter and answer the following questions.

1. Replacement gilts are selected to display the following traits:

2. Briefly explain why selecting for soundness and structural correctness in swine is vital.

3. What gene mutation causes porcine stress syndrome?

4. What differentiates swine from other livestock species like cattle, sheep, or poultry?

5. Why should replacement gilts have large feet with proportional, evenly spaced toes?

6. Most structural issues in swine are found in the _____, _____, and related joints such as the _____, _____, or _____.

7. _____ _____ is used to determine prices for market animals.

8. An upward-tilted vulva makes the animal more vulnerable to health conditions like _____ and _____.

9. Why is the swine industry able to change so quickly and effectively?

10. From the top of the shoulder to the point of the shoulder should be approximately a _____ angle.

____

11. Name the first ideal market hog unveiled by the National Pork Producers Council in 1983.

____

12. A finished animal with excess fat is considered _____.

____

13. The National Pork Board recognizes _____ major American swine breeds.

____

Key Terms

Carefully study the chapter and then use the word bank to match the term to the correct definition.

A. heritability
B. estimated breeding value (EBV)
C. terminal sire index (TSI)
D. replacement gilt
E. best linear unbiased prediction (BLUP)
F. non-productive sow days (NPD)
G. swine testing and genetic evaluation system (STAGES)
H. sow productivity index (SPI)

1. _____ The overall measure of an animal's genetic merit as a parent
2. _____ The proportion of performance differences that are due to genetics (vs. environmental influence) and is expressed as a percentage
3. _____ High-quality female hogs raised in commercial production that exhibit the correct characteristics for breeding
4. _____ Combines growth (days to 250 lbs. or DAYS), pounds of fat free lean (LBS), and back fat (BF)
5. _____ A program that calculates the most economically significant traits and predicts the genetic value of each animal
6. _____ Term used for any day that a reproductive female is not pregnant or lactating
7. _____ A statistical modeling tool that estimates the genetic merit of a particular animal and is a common on-farm tool used by producers
8. _____ An indicator of a sow's reproductive traits

Critical Thinking

Using the information learned in this chapter, read the question and provide a detailed answer.

1. Describe the importance of genetics in the different parts of the swine industry and examples of how genetics has affected the efficiency of the global industry.

____

____

____

____

Name ______________________ Date ____________ Period ________

Activity: Swine Selection and Evaluation

Practicing Swine Showmanship

Required Materials

- Laboratory notebook
- Pen or pencil
- Farm or laboratory outfit

Activity Specification

Swine showmanship is an activity performed by members of agriculture organizations such as FFA and 4-H. Today, your instructor will demonstrate the basics of showing swine. You will take notes below before being tasked with showing a barrow or gilt yourself!

NOTES:

Reflective Questions

Answer the following questions relating to the activity.

1. Analyze the behavior of swine as they are being led across the arena. What different factors might distract their performance?

2. Evaluate the body posture of the students showing the animals. What are three attributes of their posture that stand out to you?

3. Create a list of five tips for new handlers to employ when showing their swine.

Notes

Name ____________________ Date ____________ Period ________

CHAPTER 22

Classification and Breeds of Swine

Chapter Review

Carefully study the chapter and answer the following questions.

1. A(n) _____ _____ is the pig's identification from the producer.

2. This breed is black with six white points: the feet, nose, and tip of the tail.

3. The _____ breed are red and known in England as "mahogany" or "grizzly" pigs.

4. _____ is measured at the tenth rib and can be used to identify how fat or lean an animal will be.

5. This breed is black with white points and to be pedigreed, all four feet must be white with white on the face and tail.

6. Breeds that have "_____" in their name have erect ears.

7. This breed is the fifth most recorded in the United States and are referred to as "America's Sowherd."

8. The _____ breed must be red and are considered the most popular terminal breed due to their lean gain efficiency, carcass yield, and carcass quality.

9. _____ _____ indicates how many piglets the sow raised to 21 days of age.

10. This breed is often used in crossbreeding rotations, has large drooping ears, and is required to be all white for registration.

11. The _____ breed has drooping ears with the most desirable color combination being 50% black and 50% white.

12. Known as the "Old English Breed," these hogs have a white belt that can partially or fully wrap around their black bodies.

13. The sister organization of the National Swine Registry (NSR) is _____.

____

14. The _____ breed was the first to have pedigrees registered in herd books.

____

15. This breed was well known for producing large amounts of lard when they first arrived in America around 1830.

____

Key Terms

Carefully study the chapter and then use the word bank to match the term to the correct definition.

A. heritage breed
B. maternal breed
C. pedigree
D. terminal breed

1. _____ Breed that produces superior mothering traits
2. _____ Breed that produces superior carcass traits for production
3. _____ A record of the genetic makeup and line of descent of an animal, showing it to be purebred
4. _____ Breed defined by the Livestock Conservancy as being in danger of extinction

Critical Thinking

Using the information learned in this chapter, read each question and provide a detailed answer.

1. Define, in your own words, what maternal and terminal breeds of swine are and the major identifying characteristics of each.

____

____

____

____

2. List three examples of breeds that fall under terminal breeds, then do the same for maternal breeds.

____

____

____

____

____

____

Name ________________ Date ________ Period ______

Activity: Classification and Breeds of Swine

Practicing Ear Notching

Required Materials

- Laboratory notebook
- Pen or pencil
- Farm or laboratory outfit with gloves
- Dog chew pig ears
- Alcohol
- Towels
- Ear notcher

Activity Specification

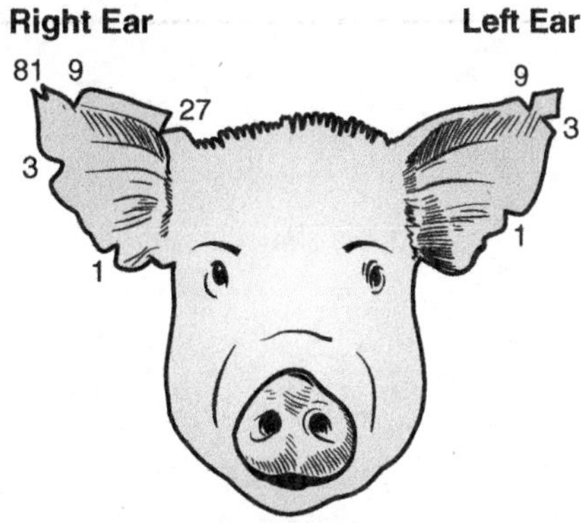

RomanYa/Shutterstock.com

Ear notching is an identification strategy that swine farmers employ to catalog and keep track of their animals. The right ear of a pig identifies the litter it came from, and the left ear identifies the individual pig number within the litter. Notching is done shortly after pigs are born, and notches grow in size as pigs grow. Within each ear, notch numbers are additive.

Your teacher will give you two pairs of dog chew pig ears. For each pair, you will be required to practice notching the appropriate litter number and pig number that are assigned to you.

Additionally, draw three pig heads in the box that represent different assigned litter and pig numbers.

1. Litter #: ________________ Pig #: ________________
2. Litter #: ________________ Pig #: ________________
3. Litter #: ________________ Pig #: ________________

Reflective Questions

Answer the following questions relating to the activity.

1. Analyze three ear-pair notches. Identify the represented litter number and pig number for each.

____

____

____

2. Evaluate the clarity of a classmate's ear notches. What recommendations would you have for future students learning to ear notch so that their notches come out clear?

____

____

____

3. Create an image below that represents a pig from farm litter 137, pig number 11.

Name ______________________ Date ____________ Period ________

CHAPTER 23 Sheep Industry Overview

Chapter Review

Carefully study the chapter and answer the following questions.

1. List the four sectors of the sheep industry.

______________________ ______________________

______________________ ______________________

2. Wool is measured in _____.

____

3. More than _____ years ago, sheep were first domesticated for human use.

____

4. Name three important organizations that help promote the sheep industry in the United States.

______________________ ______________________

____

5. The United States has approximately _____ sheep.

____

6. The first cloned animal was a sheep named _____.

____

7. What happens to wool that has dark fibers? Why?

____

____

8. List three kinds of cheese made from sheep's milk.

______________________ ______________________

____

9. _____ is the scientific name for sheep.

____

10. What caused the decline in lamb consumption in the United States after World War II?

____

____

____

11. There are an estimated _____ dairy sheep farms in the United States, mainly located in _____ and the upper Midwest.

____

____

12. Today, the US sheep industry contributes more than _____ into the economy each year.

____

Key Terms

Carefully study the chapter and then use the word bank to match the term to the correct definition.

A. ewe	E. flock	I. band
B. lamb	F. club lambs	J. mutton
C. ram	G. gregarious	
D. wether	H. laparoscopic AI	

1. _____ Sheep that are marketed to youth exhibitors who belong to organizations like National FFA or a 4-H club
2. _____ A group of 1000 sheep
3. _____ Female sheep
4. _____ Male sheep
5. _____ Meat from a sheep that is over a year old
6. _____ A minimally invasive surgical procedure that involves placing semen directly into the uterine horn of the ewe
7. _____ Young sheep (male or female), less than a year old
8. _____ Social group that makes sheep easy to move and direct when on an open range or in a pasture
9. _____ Desiring social grouping and herds
10. _____ Castrated male sheep

Critical Thinking

Using the information learned in this chapter, read the question and provide a detailed answer.

1. In this chapter, we learned why mutton consumption in the United States decreased after World War II, but in other countries, it has increased. What are some reasons that raising large flocks of sheep is more popular in countries like the United Kingdom than it is here?

____

____

____

____

____

____

Name ______________________ Date ____________ Period ________

Activity: Sheep Industry Overview

Using a Speculum

Required Materials

- Laboratory notebook
- Pen or pencil
- Farm or laboratory outfit with gloves
- Small ruminant speculum
- Lubricant
- Source of light
- Alcohol-soaked gauze
- Camera

Activity Specification

The artificial insemination of a small ruminant, such as a goat or a sheep, benefits from the use of a vaginal speculum. A vaginal speculum with applied lubricant can be used to open up the vaginal canal prior to and during insemination. This ensures a more precise delivery of sperm through the cervical interdigital prominences and into the uterus. Your instructor will demonstrate how to insert a speculum into a sheep, and you will do the same.

1. Clean the vulva in a downward motion with alcohol-soaked gauze.
2. With gloves on, apply lubricant to the vaginal speculum.
3. Insert the vaginal speculum into the restrained sheep.
4. Open the vaginal speculum and use light to identify the cervix.
5. Take two to three high-quality pictures of the cervix.
6. Insert images of the cervix in the box below and clearly label them.

Reflective Questions

Answer the following questions relating to the activity.

1. Analyze the ease of using a vaginal speculum. What technique(s) would you need to practice to improve your efficiency?

2. Evaluate the sheep's cervix, describing its appearance and consistency.

3. Create a drawing of the cervix through the speculum.

Name ______________________ Date ______________ Period __________

CHAPTER 24 Sheep Production Systems

Chapter Review

Carefully study the chapter and answer the following questions.

1. This type of disease can occur in both domestic and bighorn sheep and can cause great losses to both populations when they interact.

   ____

2. _____, _____, _____, _____, and _____ are popular breeds used in farm flock operations.

   ____

   ____

   ____

   ____

   ____

3. How are sheep production systems classified, and what are they called?

   ____

   ____

4. Most lambs sold for meat consumption are between _____ and _____ months of age.

   ____

   ____

5. Rams usually reach breeding age at around _____.

   ____

6. What are three common guard animals used in sheep operations?

   ______________________ ______________________

   ____

7. How does the sheep production cycle begin?

   ____

   ____

8. List four white-faced wool breeds commonly used in range operations.

   ______________________ ______________________

   ______________________ ______________________

9. Briefly describe why composting manure is a responsible option for producers.

   ____

   ____

10. What are some issues producers may encounter due to sheep's naturally gregarious behavior?

____

____

11. Why does the value of the carcass decrease when a lamb matures to a yearling, and how is a lamb carcass distinguished from a yearling?

____

____

12. Ewes are typically bred when they are _____ to _____ months old.

____

____

13. How long does the gestation period for sheep typically last?

____

Key Terms

Carefully study the chapter and then use the word bank to match the term to the correct definition.

A. band	D. lambing jug	G. shearing
B. farm flock operation	E. niche market	H. tub system
C. lamb feeder operation	F. range operation	

1. _____ A covered pen used for lambing
2. _____ Operation that generally has between 30 and 200 head of sheep in production
3. _____ Operation that purchases lambs at weaning or around 80 pounds to feed until they are ready for market or the market price increases
4. _____ The removal of the wool fleece with cutters
5. _____ A specialized, targeted marketing strategy in which a producer aims to meet the wants of a small but dedicated consumer base
6. _____ Group of 1000 ewes and lambs that are grazed together
7. _____ Used to gather a group of sheep and funnel them into a single-file chute system
8. _____ Operation that is prevalent in Western states due to the large amount of grazing land

Critical Thinking

Using the information learned in this chapter, read the question and provide a detailed answer.

1. Describe the differences between range flocks and farm flocks in your own words, including differences in the breeds used and why they fit one system better than the other.

____

____

____

____

Name ______________________ Date ______________ Period __________

Activity: Sheep Production Systems

Flipping a Sheep

Required Materials

- Laboratory notebook
- Pen or pencil
- Farm or laboratory outfit with gloves

Activity Specification

One of the most fundamental handling strategies in working with sheep is flipping them. Flipping a sheep involves having the animal sit on its rump with its limbs splayed forward as the back of the animal rests on your legs. Flipping a sheep is an important strategy that can be used to shear sheep, administer medicine, or even to simply do a physical check-up.

Air Images/Shutterstock.com

Your instructor will demonstrate how to flip a sheep and you will replicate this same procedure.

Reflective Questions

Answer the following questions relating to the activity.

1. Watch your classmates flipping sheep. What is something that you note from this perspective that you can implement to improve your own flipping technique?

2. Evaluate the behavior of sheep when they are flipped. What can be done to ensure they remain calm so procedures can be completed?

3. Create a four-step drawing that demonstrates how sheep can be flipped and a procedure that can be done when they are flipped.

Name ______________________ Date ____________ Period ________

CHAPTER 25 Sheep Management and Care

Chapter Review

Carefully study the chapter and answer the following questions.

1. Balanced feeds for sheep have four main components.

2. Why is it important to increase the amount of grain provided to ewes during the last 45 days of gestation?

3. What infection is not especially common in flocks, due to the amount of wool and lanolin they regularly display, compared to club lambs who are regularly shorn?

4. A maintenance feed ration for older ewes may contain _____ protein.

5. Body condition scores for sheep range from _____ to _____.

6. Teeth size, shape, and spacing will change over the life of a sheep and may start to wear out after _____ _____.

7. This illness is also called "contagious ecthyma" or "orf" and is a viral skin disease.

8. What type of identification can be used to send carcass merit information from lamb harvest facilities back to the producer?

9. _____ is a common fat supplement for sheep.

10. As ruminants, sheep will achieve optimal health if they have _____ in their diet.

11. _____ and _____ are high in protein.

12. This genetic disorder is caused by a recessive gene that causes lambs to be born with skeletal and muscular deformities.

13. This bacterial infection is found in the hoof and is extremely difficult to eradicate due to the bacteria being spread in muddy soils and bedding.

____

14. Vitamins _____, _____, and _____ should be fed to sheep through supplements.

____

____

____

15. Ewes typically lamb within _____ to _____ minutes.

____

____

Key Terms

Carefully study the chapter and then use the word bank to match the term to the correct definition.

A. breeding harness	D. teaser	G. nutritional flushing
B. homozygosity	E. line breeding	H. pure (straight) breeding
C. inbreeding	F. marking crayon	I. outbreeding

1. _____ Used for identifying sheep that have been treated with medication or need to be separated from the flock
2. _____ Clarified dominant and recessive genes that may result in lowered reproductive performance and growth
3. _____ Apparatus for rams that holds a crayon to mark females who have been mounted
4. _____ When a producer uses unrelated rams of the same breed to breed his ewes
5. _____ Mating two animals of the same breed
6. _____ A ram that has been vasectomized and will not be able to impregnate ewes
7. _____ A specialized type of inbreeding that concentrates mating a specific individual
8. _____ The mating of rams and ewes that have a common ancestor
9. _____ The process of increasing feed quality for a ewe for the three to four weeks prior to breeding

Critical Thinking

Using the information learned in this chapter, read each question and provide a detailed answer.

1. Select an illness found in sheep that was covered in this chapter and interests you. What are the symptoms that a producer should look out for, and how is the illness transmitted?

____

____

____

2. What components of a herd health plan are particularly important to preventing this illness? Why?

____

____

____

____

Name ______________________ Date ______________ Period __________

Activity: Sheep Management and Care

Shearing Sheep

Required Materials

- Laboratory notebook
- Pen or pencil
- Farm or laboratory outfit
- Electric sheep clippers

Activity Specification

Now that you have mastered the art of flipping sheep, you will learn how to shear them. Shearing sheep is one of the central pillars of the sheep industry. Shearing is done to collect wool. It must be done swiftly and carefully to not compromise the integrity of the wool or to stress the sheep more than is necessary.

Your instructor will demonstrate how to shear sheep and then you will do the same.

Reflective Questions

Answer the following questions relating to the activity.

1. Analyze the surroundings where you were clipping sheep. What safety hazards exist?

____

____

____

____

2. Evaluate how well you did with your clipping. What are two ways that you can improve your clipping?

____

____

____

____

3. List at least four different locations on a sheep where you must be especially careful when clipping.

____

____

____

____

Notes

Name ______________________ Date ______________ Period __________

CHAPTER 26

Sheep Selection and Evaluation

Carefully study the chapter and answer the following questions.

1. Characteristics to consider when evaluating breeding ewes include:

_______________ _______________

____

2. In a breeding sheep show, the sheep are typically classified by breed and then grouped into _____ by _____.

____

____

3. When wool is a product or by-product of your operation, you should consider a dense fleece with a distinct _____ that is free of _____ _____.

____

____

____

4. Briefly explain why structural correctness is important, no matter the purpose a lamb is intended to meet.

____

____

5. The ideal market lamb weighs between _____ and _____ pounds.

____

____

6. List the three main considerations when evaluating and selecting market lambs.

_______________ _______________

____

7. What does the presence of good finish indicate?

____

8. In a breeding sheep show, why are the sheep sorted by age?

____

9. _____ lambs suffer from extra strain on joints and soreness or lameness.

____

10. The ideal market lamb should display the following characteristics:

_______________ _______________

_______________ _______________

11. A _____ and _____ rump is preferred in market lambs.

____

____

12. List the three highest-value cuts of meat taken from a lamb.

______________________ ______________________

____

Key Terms

Carefully study the chapter and then use the word bank to match the term to the correct definition.

A. frame size
B. balance
C. hindsaddle
D. cow-hocked
E. sickle-hocked
F. finish
G. monkey mouth
H. parrot mouth
I. National Sheep Improvement Program (NSIP)
J. muscling

1. _____ Too much set or angle to the hock
2. _____ The proportion of body parts
3. _____ Degree of muscle or thickness
4. _____ Lower jaw extending further forward than the upper jaw
5. _____ Amount of external fat on the lamb
6. _____ Collects data points on breeding sheep and their offspring to provide information about heritable characteristics to breeders
7. _____ Indicates the approximate weight of the lamb when it is finished
8. _____ Upper jaw extending further forward than the lower jaw
9. _____ From the last rib to the dock on the sheep
10. _____ Hocks that are pushed out away from the body

Critical Thinking

Using the information learned in this chapter, read the question and provide a detailed answer.

1. How do programs like the National Sheep Improvement Program and other similar livestock species programs assist their industries in the growth and production of quality animal protein and products?

____

____

____

____

____

____

____

____

____

____

Name ______________________ Date ____________ Period ________

Activity: Sheep Selection and Evaluation

Sheep Showmanship

Required Materials

- Laboratory notebook
- Pen or pencil
- Farm or laboratory outfit
- Halter

Activity Specification

Sheep showmanship is an activity performed by members of agriculture organizations such as FFA and 4-H. Today, your instructor will demonstrate the basics of showing sheep. You will take notes below before being tasked with showing a sheep yourself!

NOTES:

Reflective Questions

Answer the following questions relating to the activity.

1. Analyze the behavior of sheep as they are led across the arena. What are three different factors that might distract their performance?

2. Evaluate the body posture of the student showing the animal. What attributes of their posture stand out to you?

3. Create a list of five tips that new handlers can employ when showing their sheep.

Notes

Name ____________________ Date __________ Period __________

CHAPTER 27 Classification and Breeds of Sheep

Carefully study the chapter and answer the following questions.

1. This Spanish breed was developed for its superior fleeces and has become the basis for all fine-wool breeds with exceptionally fine, soft fleeces that command high prices around the world.

2. Considered to be the world's most prolific dairy sheep, this breed has a "rat tail" that is narrow and does not grow wool.

3. Developed in England, this breed grows quickly and produces muscular carcasses with high yield grades that make them common in purebred and crossbred operations.

4. This breed was developed using Rambouillet rams and Rambouillet/Corriedale/Lincoln ewes in Dubois, Idaho.

5. This breed was developed in Maine, is hardy and adaptable to most environments in the United States, and does not produce wool.

6. There are estimated to be more than _____ breeds of sheep in the world.

7. _____ is often cited as the most influential person in developing breeds of sheep for meat production and carcass quality, now known as _____ breeds.

8. This heritage breed can be multi-horned and is colored with large light and dark patches.

9. This large breed of sheep has a solid white fleece, a black face and ears, black legs, and is often used to produce club lambs.

10. Sheep of this breed are solid white with pink noses, white hooves, and well-muscled carcasses.

____

____

11. There are more than _____ breeds of sheep in the United States.

____

12. This breed is commonly seen in a mousy-gray color, typically has two to four lambs at a time, and was developed in Finland.

____

____

13. Sheep breeds can be classified based on characteristics such as:

______________________ ______________________

____

Key Terms

Carefully study the chapter and then use the word bank to match the term to the correct definition.

A. ewe breeds
B. fine-wool breeds
C. hair sheep
D. ram breeds
E. open-faced
F. long-wool breeds
G. medium-wool breeds
H. dual-purpose breeds
I. piebald
J. polycerate

1. _____ Wool does not grow on or around the sheep's eyes and face
2. _____ Produce long, coarse fiber that is well-suited for hand-spinning
3. _____ Breeds that are mainly used for producing meat with fleece as a secondary concern
4. _____ White-faced breeds with superior mothering abilities
5. _____ Polyestrous breeds that tend to be more parasite-resistant and heat-tolerant
6. _____ Multi-horned
7. _____ Sheep that produce strong maternal and terminal characteristics
8. _____ Used for terminal sires; common in meat production operations
9. _____ Produce wool with tightly crimped, small-diameter fibers
10. _____ Irregular patches of color

Critical Thinking

Using the information learned in this chapter, read the question and provide a detailed answer.

1. Why is it important that different breeds of sheep are produced? What purposes do such diverse production operations serve?

____

____

____

____

____

Name ______________________ Date ____________ Period ________

Activity: Sheep Classification and Breeds

Practicing Sheep Castration Technique

Required Materials

- Laboratory notebook
- Pen or pencil
- Farm or laboratory outfit with gloves
- Elastrator
- Bands
- Sock with two dates

Activity Specification

Castration is a common procedure in livestock operations. It reduces unwanted behaviors in animals and protects the flavor of meat produced. Castration can involve surgery, or it can be bloodless. Today you will practice the technique used in bloodless castration.

Procedure

1. Place two dates into a sock.
2. Arm an elastrator with a band.
3. Using the elastrator, carefully open the band and pass the band over the two dates in the sock.
4. Slowly begin to close the elastrator and with your finger flick the band from the elastrator spokes to apply it to the sock.

Reflective Questions

Answer the following questions relating to the activity.

1. Analyze the integrity of the elastrator band. Why is it important for farmers to make sure that bands are still in place after they have been applied?

____

____

____

2. Evaluate the ease of applying a band around the "testicles" of your model. In a real castration scenario, what other safety factors would you have to account for?

____

____

____

3. Compare and contrast castrating animals via elastrators versus using surgical methods.

____

____

____

Notes

Name ________________ Date ________ Period ______

CHAPTER 28 Goat Industry Overview

Carefully study the chapter and answer the following questions.

1. As of June 2020, there were _____ goats living in the United States.

 ____

2. The _____ breed is the primary type of meat goat in the United States.

 ____

3. Dairy goats were first brought over by _____ settlers in the _____ and _____ centuries.

 ____

 ____

4. Goat milk is known to have the same nutritional value as cow's milk, but with _____ _____.

 ____

 ____

5. _____ is the leading country in modern goat production, with a population of more than _____ goats.

 ____

 ____

6. List the four main sectors of the goat industry.

 ______________ ______________

 ______________ ______________

7. The average milking doe will produce between _____ and _____ pounds of fluid milk per day over the course of her _____ day lactation.

 ____

 ____

 ____

8. It is most common for goats to be harvested for meat when they are young, weighing between _____ and _____ pounds.

 ____

 ____

9. Angora goats were imported into the United States from Turkey for _____ production in 1849.

 ____

10. List the three main organizations that help promote the US goat industry.

 ______________ ______________

 ____

11. _____ is the state with the most goat production.

12. As of 2022, there were approximately _____ milk goats and kids in the United States.

13. What is the scientific name for the goat species?

14. How does goat meat produced by younger animals differ from the meat produced by older animals?

15. The United States imports _____% of the goat meat consumed here each year.

Key Terms

Carefully study the chapter and then use the word bank to match the term to the correct definition.

A. billy
B. buck
C. cashmere
D. chevon
E. disbudding
F. doe
G. kid
H. kidding
I. nanny
J. targeted grazing
K. wether

1. _____ Rarely used term for goat meat
2. _____ Made from the soft fibers of the undercoat that many goat breeds produce during cold seasons
3. _____ Castrated male goat
4. _____ Historical name for a female goat
5. _____ Female goat of any age
6. _____ This practice uses goats to go into overgrown areas and consume natural forage, grasses, shrubs, and the like before they turn into dry brush
7. _____ Removal of horns from young kids
8. _____ Young goat (male or female), typically less than one year old
9. _____ Act of giving birth to goats
10. _____ Historical term for a male goat
11. _____ Intact male goat of reproductive age

Critical Thinking

Using the information learned in this chapter, read the question and provide a detailed answer.

1. Why is the goat industry classified as a niche industry in the United States?

Name ______ Date ______ Period ______

Activity: The Goat Industry

Goat Cheese Taste Test

Required Materials

- 6 types of goat cheese
- Eating utensils
- Water and saltine crackers
- Pen or pencil

Activity Specification

Today you will blindly taste and rank different goat cheese samples. You will receive samples and must rank them in order of your preference. The following code names have been given to the cheeses: red, orange, yellow, green, blue, violet.

1. **Sample: Red**

   A. **Description:** ______

   B. **Final Ranking:** ______

2. **Sample: Orange**

   A. **Description:** ______

   B. **Final Ranking:** ______

3. **Sample: Yellow**

   A. **Description:** ______

   B. **Final Ranking:** ______

4. **Sample: Green**

   A. **Description:** ______

   B. **Final Ranking:** ______

5. **Sample: Blue**

   A. **Description:** ______________________________

   B. **Final Ranking:** ______________________________

6. **Sample: Violet**

   A. **Description:** ______________________________

   B. **Final Ranking:** ______________________________

Reflective Questions

Answer the following questions relating to the activity.

1. Analyze the appearance of the cheeses. What were your preconceptions of the cheeses' appearances before you tasted them?

2. Evaluate the smells of the cheeses presented. How did the smells of the cheeses correlate with your preferences?

3. Create a statement that relates the appearance, smell, and taste of the cheeses to your personal preferences.

Name ______________________________ Date ______________ Period ________

CHAPTER 29 Goat Production Systems

Carefully study the chapter and answer the following questions.

1. Briefly describe what classifies a goat operation as a small flock.

 ____

 ____

2. Most producers choose to wean kids from does when they are between _____ and _____ days old.

 ____

 ____

3. A goat's instinct to climb can be encouraged by the addition of what items to its housing?

 ____

4. Goats will select up to _____% of their diets from brush-type plants and woody shrubs, rather than grasses or perennial legumes.

 ____

5. What type of operation specializes in raising high-quality specimens of single goat species to sell to other producers?

 ____

6. Many goat breeds are _____ _____, meaning they are only able to reproduce during specific times of the year.

 ____

 ____

7. What makes fighting or aggressive behavior common across all goat breeds, genders, ages, and types?

 ____

8. Briefly describe what classifies a goat operation as a backyard operation.

 ____

 ____

9. What natural instincts can lead goats to get into troublesome situations?

 ____

 ____

 ____

 ____

10. Once a doe is pregnant, she will carry her offspring for an average of _____ days before birth.

 ____

Key Terms

Carefully study the chapter and then use the word bank to match the term to the correct definition.

A. browse
B. fenceline weaning
C. milking parlor
D. stanchion

1. _____ A room or building that is dedicated to the collection of milk from does and milk storage
2. _____ Thicker and more fibrous forage material which goats prefer over other ruminants
3. _____ A restraint made from wood or metal that generally elevates a goat about two feet above the ground
4. _____ Common practice which allows the mothers to have contact with their kids along a wire-type fence, but prevents nursing

Critical Thinking

Using the information learned in this chapter, read the question and provide a detailed answer.

1. How would you describe the different types of goat production operations? What similarities and differences are found across different types of operations?

Name ______________________________ Date ______________ Period __________

Activity: Goat Production Systems

Milking a Goat

Required Materials

- Laboratory notebook
- Pen or pencil
- Farm or laboratory outfit with gloves
- Bucket
- Pre-dip and post-dip solutions dip cups
- Clean water
- Clean towels
- Dairy goat

Activity Specification

Today your instructor will demonstrate the proper commercial procedure for milking animals. Modern intensive dairies involve many automated procedures. However, milk technicians are always needed to address machine-error and ensure proper milking. Your instructor will demonstrate proper milking procedure for milking dairy goats and then you will be expected to demonstrate mastery of the procedure yourself.

1. Clean the goat's udder using warm water with a very mild detergent. Use a towel to remove any debris, and discard the towel when you are done.
2. Rinse the goat's udder with warm water to remove any residual detergent.
3. Apply pre-dip solution to the teat containing iodine using the appropriate dip cup.
4. Allow pre-dip solution to remain on teat for around 30 seconds.
5. Use towel to wipe off excess pre-dip solution. Discard the towel.
6. Milk the goat.
7. Apply post-dip solution to the teat using the appropriate dip cup.
8. Allow post-dip solution to remain on teat for at least 30 seconds. Allow the teat to air dry.

Reflective Questions

Answer the following questions relating to the activity.

1. Analyze the procedure. Why were so many steps taken to ensure the cleanliness of the goat mammary gland?

____

____

____

____

2. Evaluate your success in obtaining milk. What could you have done differently to improve milk collection?

____

____

____

____

3. Create an image that illustrates the inside of a teat, and how squeezing from the top down is necessary to achieve successful expulsion.

Name ______________________ Date ______________ Period __________

CHAPTER 30 Goat Management and Care

Carefully study the chapter and answer the following questions.

1. This illness presents as pus-filled abscesses near lymph glands and can be treated with a vaccine.

____

2. Goats are _____, and they need to consume a broad range of forage types to maintain their digestive health.

____

3. What are three reasons that kids need colostrum in the early hours of their lives?

____

____

____

4. How often do goats kept in dry lots, pens, or pasture need their hooves trimmed?

____

5. List five methods used to identify goats.

______________________ ______________________

______________________ ______________________

____

6. This disease is also known as paratuberculosis and is caused by a mycobacterium that builds up in the gastrointestinal tract.

____

7. For goats, _____% of fetal growth occurs in the final _____ weeks of pregnancy.

____

____

8. Goats do not have _____ _____ in their skin, and often pant to maintain temperature during heat stress.

____

____

9. List the minerals that goats may need to have supplemented in their feed.

______________________ ______________________

______________________ ______________________

____

10. This disease is caused by a common lentivirus that slowly progresses after infection, and in young kids will cause paralysis and seizures.

____

11. Feed can cost _____% of the total cost of goat production.

____

12. This disease is caused by a virus that presents as lesions around the mouth and on the feet, teats, or other parts of the goat.

____

13. The estrous cycle of a doe takes an average of _____ days.

____

Key Terms

Carefully study the chapter and then use the word bank to match the term to the correct definition.

A. bicornuate
B. caruncles
C. creep
D. ketosis
E. esophageal groove
F. linear appraisal system
G. magnesium
H. selenium
I. copper
J. iron

1. _____ Uses distinct anatomical locations to determine the amount of condition on the goat
2. _____ Also known as toxemia
3. _____ Supports tissue growth and other functions throughout the body
4. _____ Required micromineral that may be deficient in some soils and native plants
5. _____ A uterus with two distinct horns
6. _____ A muscular passageway that nursing kids use to ingest milk directly into the abomasum
7. _____ Required for rapidly growing animals and may be deficient in some soil types
8. _____ A feeding area for kids set up away from their mothers
9. _____ Small round structures in the doe's uterine wall
10. _____ Needed to maintain metabolic health

Critical Thinking

Using the information learned in this chapter, read the question and provide a detailed answer.

1. What does it take to maintain a healthy goat herd? Describe the steps that producers can and should take to protect biosecurity in their herds, and preventative care or management measures that should be taken to keep goats healthy.

____

____

____

____

____

____

____

____

____

Name ______________________ Date ______________ Period __________

Activity: Goat Management and Care

Trimming Goat Horns

Required Materials

- Laboratory notebook
- Pen or pencil
- Farm or laboratory outfit with gloves
- Horn trimmers
- Horns

Activity Specification

Today you will get hands-on experience with the amount of force needed to trim goat horns. A horn trimmer is a tool used to trim down or clip horns that have already grown in. Though other methods can be used to prevent horn growth or trim horns, this method is quick and can be done with a restrained goat.

A. Your instructor will hold a horn in their hands.

B. You will use the horn trimmers to clip off about ½ inch off the horn. Do this swiftly and recognize the amount of force that is needed to successfully clip the horn.

Reflective Questions

Answer the following questions relating to the activity.

1. Analyze the consistency of the horn you are clipping. Explain its appearance and texture.

____

____

____

2. Evaluate the smoothness of the clipping you did. How can it be improved?

____

____

____

3. Create a list of three safety hazards that can result from performing the clipping procedure on goats.

____

____

____

How could you address these hazards?

____

____

____

Notes

Name ______________________ Date ____________ Period ________

CHAPTER 31 Goat Selection and Evaluation

Carefully study the chapter and answer the following questions.

1. The following characteristics are priorities when selecting meat goats for show:

______________________ ______________________

______________________ ______________________

2. When evaluating dairy goat females for production, characteristics to consider include:

______________________ ______________________

______________________ ______________________

3. _____ refers to the lower jaw being further forward than the upper jaw.

____

4. The udder on a dairy goat should be _____ and _____ when viewed from the side and rear.

____

____

5. An ideal meat goat that is ready for harvest will weigh between _____ and _____ pounds.

____

____

6. Why is it important to cull breeding bucks with cryptorchidism from the herd?

____

____

7. What characteristics are important to consider when selecting breeding does?

____

____

____

8. The _____ collects data points on breeding sheep and goats and their offspring to provide information on heritable characteristics to breeders.

____

9. How would you describe an ideal dairy goat's teats?

____

____

10. The _____ and _____ of a goat are the two main areas used to check skeletal correctness when selecting goats.

____

____

11. What characteristics make a doe more feminine?

____

____

12. Each doe in a dairy herd must exhibit traits that will allow her to:

______________________ ______________________

____

Key Terms

Carefully study the chapter and then use the word bank to match the term to the correct definition.

A. cryptorchidism
B. evaluate
C. selection
D. body capacity
E. dairy character
F. finish
G. splayfooted
H. post-legged
I. pigeon-toed
J. cow-hocked

1. _____ The presence of subcutaneous fat
2. _____ Having feet turned out
3. _____ To make a judgment about the value, quality, or nature of something
4. _____ Condition characterized by the presence of undescended testicles
5. _____ Having feet turned in
6. _____ Term used to evaluate increased barrel depth and width of the chest
7. _____ A producer's careful evaluation and addition of animals to a flock that bring desired characteristics
8. _____ Having hocks that nearly touch
9. _____ Having legs that are too straight up and down and set too far under the body
10. _____ Characteristics that make goats look like milk producers

Critical Thinking

Using the information learned in this chapter, read the question and provide a detailed answer.

1. If you are looking to bring breeding bucks into a dairy herd, what characteristics might you consider? Why are these characteristics especially important?

____

____

____

____

____

____

____

____

Name ______________________ Date ______________ Period __________

Activity: Goat Selection and Evaluation

Goat Showmanship

Required Materials

- Laboratory notebook
- Pen or pencil
- Farm or laboratory outfit
- Halter

Activity Specification

Goat showmanship is an activity performed by members of agriculture organizations such as FFA and 4-H. Today, your instructor will demonstrate the basics of showing goats. You will take notes below before being tasked with showing a goat yourself.

NOTES:

Reflective Questions

Answer the following questions relating to the activity.

1. Analyze the behavior of goats as they are being led across the arena. What factors seem to distract their performance?

2. Evaluate the body posture of the students showing goats. What are three attributes of their posture that stand out to you?

3. Create a list of five tips that new handlers can employ when showing their goats.

Name ________________________ Date ____________ Period ________

CHAPTER 32 Classification and Breeds of Goats

Carefully study the chapter and answer the following questions.

1. The _____ is one of the oldest goat breeds, originating in central Turkey, with roots that can be traced back to biblical times.

   ____

2. This breed is the newest to be recognized by the ADGA and has a golden body color with hair that can range from short to long and flowing.

   ____

3. There are estimated to be about _____ breeds of goats in the world.

   ____

4. The _____ breed is known for meat production and was first developed in the early 1900s in South Africa.

   ____

5. This distinctive breed is known for its unique look and is unable to be identified with ear tags.

   ____

6. The most common goat breeds were brought to the United States from _____.

   ____

7. This goat breed has white hair and black skin, droopy ears, and a stocky, strong body.

   ____

8. The _____ is an organization dedicated to representing the interests of domestic goat producers.

   ____

9. The _____ breed has long, flopping ears and produces milk that is high in butterfat and protein.

   ____

10. The _____ breed was developed in New Zealand with the goal of producing a hardy animal that gained weight quickly, but can have an aggressive temperament.

    ____

11. The _____ advocates for the interests of dairy goat breeds in the United States.

    ____

12. This breed's mature weight is anywhere between 60 and 85 pounds and can have colors ranging from light to dark caramel brown, solid black or gray, and black/gray agouti.

    ____

13. This breed is the only miniature breed recognized by the ADGA and can display most any color pattern.

    ____

Key Terms

Carefully study the chapter and then use the word bank to match the term to the correct definition.

A. agouti C. pinna E. kiko
B. mohair D. wattle

1. _____ External ear cartilage
2. _____ Long fleece of Angora goats
3. _____ A small appendage on a goat's neck
4. _____ Maori word for "meat"
5. _____ Occurs when individual hair fibers have multiple colors

Critical Thinking

Using the information learned in this chapter, read the question and provide a detailed answer.

1. What traits separate meat goat breeds from dairy goat breeds? If you were not sure what purpose a breed of goat was supposed to serve, how would you go about figuring it out?

____
____
____
____
____
____
____
____
____

Name ______________________ Date ______________ Period __________

Activity: Classification and Breeds of Goats

Presenting Goat Breeds

Required Materials

- Computer with internet access
- Gmail account

Activity Specification

Today you will create a PowerPoint®/Google Slides presentation about goat breeds, showcasing at least six different breeds.

Your PowerPoint deck should:

- Include 10 slides in total
  - Include a title page and a reference page
- Introduce the scientific name of the animal and any related scientific terms (age/sex)
- Describe the average ages of puberty and breeding and length of gestation
- Cover three dairy goat breeds and three meat goat breeds
  - Give breed details and their purposes

When you have completed your slides, be prepared to give a presentation to the class.

Reflective Questions

Answer the following questions relating to the activity.

1. Analyze your classmates' presentations. Is there a common theme regarding the most common breed of goat covered for either meat or milk?

____

____

____

2. Evaluate your own presentation and indicate at least two aspects where you could have improved. What would you change if you did it again?

____

____

____

3. Create a picture of what your ideal goat breed would look like and describe three traits that would make this goat breed ideal.

Name ______________________ Date ______________ Period __________

CHAPTER 33

Dairy Industry Overview

Carefully study the chapter and answer the following questions.

1. Give four examples of common processed dairy products.

_______________ _______________

_______________ _______________

2. The dairy cattle industry provides more than $_____ in direct economic impact to the United States each year.

____

3. _____ is the French chemist responsible for developing a process for heating milk to remove pathogens called _____.

____

____

4. Most estimates are that somewhere between _____% and _____% of the beef processed in the United States comes from dairy cattle.

_______________ _______________

5. How did milk maids play a role in the development of the smallpox vaccine and important advancements in immunology? Who made this discovery?

____

____

____

6. List five organizations that promote the dairy industry in the United States.

_______________ _______________

_______________ _______________

____

7. Dairy cattle in the United States produce more than _____ gallons of milk annually.

____

8. How many head of dairy cattle are the following countries home to?

A. India: _______________

B. Brazil: _______________

C. United States: _______________

9. List four ways that today's dairy farms maximize efficiency.

_______________ _______________

_______________ _______________

10. A single mature dairy cow produces between _____ and _____ gallons of milk per day.

_______________ _______________

11. Animals within the family *Bovidae* are large cloven-hooved animals with ruminant digestive systems including _____, _____, _____, and _____.

____

____ ____

12. The dairy cattle industry directly employs about _____ Americans.

____

13. List three examples of support services for the dairy industry.

____ ____

____

14. There is evidence of people consuming cow's milk as a portion of their diet as early as _____.

____

15. The top five dairy cattle states by population are.

____ ____

____ ____

____

Key Terms

Carefully study the chapter and then use the word bank to match the term to the correct definition.

A. calving	D. cow	G. bull calf	J. steer
B. heifer	E. calf	H. Sumerians	
C. dry cow	F. bull	I. springer	

1. _____ Intact male dairy animal of reproductive age
2. _____ The act of giving birth to cattle
3. _____ Dairy cow that is close to calving
4. _____ Uncastrated male dairy animal that is two years old or younger
5. _____ Young female dairy animal that has not calved
6. _____ Castrated male dairy animal
7. _____ Mature female cow kept in dairy operations for breeding
8. _____ Young cattle (male or female), typically less than six months of age
9. _____ Mature female dairy animal that has calved
10. _____ The first people to raise domesticated cattle for milk

Critical Thinking

Using the information learned in this chapter, read the question and provide a detailed answer.

1. Describe the different sectors of the dairy industry. Is it a more complicated industry than you previously imagined? If it is, how so?

____

____

____

Name ____________________ Date ____________ Period ________

Activity: The Dairy Cattle Industry

Making Ice Cream

Required Materials

- 1 cup whipping cream
- 1 cup milk
- ½ cup sugar
- 1 teaspoon of flavoring (ex. vanilla extract)
- 3 cups of ice
- 1 cup of rock salt
- 2 gallon-size resealable plastic bags
- 2 spoons

Activity Specification

Today you will make ice cream with items you can likely find in your kitchen at home!

1. In a gallon-sized resealable bag, mix the whipping cream, milk, sugar, and flavoring. This is the ice cream mixture.
2. Push air out of the bag and then seal it.
3. In the other gallon-sized resealable bag, place 3 cups of ice and 1 cup of rock salt. This is the cold bag.
4. Place the ice cream mixture bag within the cold bag and carefully close the cold bag while pressing the air out.
5. For 10 minutes, shake the bags carefully.
6. Carefully remove the ice cream mixture from the cold bag and enjoy the ice cream with a friend!

Reflective Questions

Answer the following questions relating to the activity.

1. Analyze the texture of the ice cream. What could be improved for it to feel more like store-bought ice cream?

____

____

____

____

2. Evaluate the taste of the ice cream. How do you think the recipe could be improved to produce a more popular flavor?

____

____

____

____

3. Draw a dairy farm. Label the milking parlor and cooling tanks. Show a truck which transfers milk, and add a facility where milk can be pasteurized and homogenized before being made into ice cream.

4. Define *pasteurization*.

5. Define *homogenization*.

6. Salt was used in this experiment because it helps to melt ice. It does this by interfering with the solid structure of frozen water. Meanwhile, because milk is mostly water, the water inside of the ice cream mixture (which never comes in contact with the salt) begins to freeze. Using this information, explain how ice cream is made in a manner that an elementary school student could understand. Your instructor will ask you to present this explanation to get full credit for today's lab.

Name ______________________ Date ______________ Period __________

CHAPTER 34 Dairy Production Systems

Carefully study the chapter and answer the following questions.

1. In what period of the lactation cycle is milk production highest?

   ____

2. As of 2019, more than _____% of Holstein cows in the United States could trace their lineage back to _____ bulls.

   ____

   ____

3. _____ milking parlors allow cows to enter down both sides of the milking floor through side-opening gates.

   ____

4. Why is colostrum important for a newborn calf's health?

   ____

   ____

5. Describe the following:

   A. Class I milk: ______________________________

   B. Class II milk: ______________________________

   C. Class III milk: ______________________________

   D. Class IV milk: ______________________________

6. _____ is meat from cattle that are milk-fed for their entire lives.

   ____

7. _____ is a system of dairy farming in which a large portion of the forage fed to cows comes from grazing on open pasture.

   ____

8. _____ parlors bring groups of cows to each side of the milking pit and the cows stand nose to tail as they are being milked.

   ____

9. Describe the purpose of the Capper-Volstead Act of 1922.

   ____

   ____

10. Heifer calves are typically _____ to _____ months old before they go into dairy production.

    ____

11. _____ systems raise animals using mixed rations rather than grazing.

    ____

12. Most cows continue in the production cycle until their milk production begins to decline, at between _____ and _____ years old.

____

13. List the four stages of the lactation curve, in order.

_______________________ _______________________

_______________________ _______________________

14. To produce dairy calves, cows who have superior _____ qualities are bred to bulls who exhibit superior _____ qualities.

____

____

15. Because of proximity and ease of milking, some intensive systems milk cows up to _____ times per day.

____

Key Terms

Carefully study the chapter and then use the word bank to match the term to the correct definition.

A. colostrum	D. cooperative	G. lactation curve	I. milking unit
B. anaerobic lagoon	E. lactation	H. Standard Milk Ordinance	J. Federal Milk Marketing Order
C. aerobic lagoon	F. manure slurry		

1. _____ Large holding areas for wastewater treatment, where microorganisms use oxygen dissolved in the water through aeration systems to degrade organic matter into carbon dioxide
2. _____ Program that supports the prices that US dairy producers receive for their products
3. _____ Nutrient-rich first milk
4. _____ Wastewater and runoff from pens and milkhouse floors
5. _____ Milk production
6. _____ Group of producers who work together to move product to processors, selling for a consistent price
7. _____ Four stages that cows experience in milk production
8. _____ It is placed on the udder to collect milk
9. _____ Sets the guidelines that must be followed to receive permits to sell graded milk
10. _____ Large holding areas for wastewater treatment that are not aerated

Critical Thinking

Using the information learned in this chapter, read the question and provide a detailed answer.

1. What are the advantages and disadvantages of the types of milking parlors described in this chapter? How would you compare them?

____

____

____

____

____

Name ________________________ Date ____________ Period ________

Activity: Dairy Production Systems

Thawing Semen Straws for Artificial Insemination (AI)

Required Materials

- Liquid nitrogen storage tank
- Semen straws (empty)
- Water bath
- Thermometer
- AI rod
- AI rod sheath
- Scissors
- Gloves
- Tongs

Activity Specification

Today you will learn and demonstrate the proper procedure needed to thaw semen straws for artificial insemination. This is done regularly in the dairy industry. Empty straws will be used, as this will just be for handling practice.

1. Put on appropriate gloves and PPE.
2. Open the liquid nitrogen storage tank.
3. Using tongs, carefully remove one semen straw from the storage tank.
4. Place the frozen semen straw into a water bath that is no hotter than 95°F for 30 seconds.
5. Carefully use scissors to cut one end of the semen straw.
6. While holding the cut end of the semen straw upward, load it onto the AI rod without placing any pressure on the plunger.
7. Place the loaded AI rod into the AI rod sheath, ensuring the plunger is not depressed.

Reflective Questions

Answer the following questions relating to the activity.

1. What aspect of preparing an AI rod do you need to improve on? How does timeliness affect your success in each step?

   ____

   ____

   ____

   ____

2. Cleanliness is essential in this procedure to ensure the highest probability of a successful insemination. Evaluate aspects of this procedure where cleanliness could be improved.

   ____

   ____

   ____

   ____

3. Draw pictures below of each tool that was needed to set up an AI rod and clearly label them.

Name ______________________ Date ____________ Period ________

CHAPTER 35 Dairy Cattle Management and Care

Carefully study the chapter and answer the following questions.

1. _____ is one of the most serious diseases seen in dairy calves and is responsible for up to 11% of all dairy mortalities.

____

2. Gestation in dairy cows lasts between _____ and _____ days.

______________ ______________

3. Dairy cows typically have more visible _____ than other livestock.

____

4. Some breeds, like Jerseys, are more likely to suffer from _____ _____ than others because of differences in their calcium uptake receptors.

____

5. The most common parasites afflicting dairy cattle include:

______________ ______________

______________ ______________

____

6. Dairy cows are typically given free access to their feed for _____ hours per day.

____

7. Describe why thick bedding is important for dairy cattle.

____

____

8. Signs of impending birth in dairy cattle include:

______________ ______________

______________ ______________

______________ ______________

____

9. _____ is caused by a subspecies of the *Mycobacterium avium* bacteria and is sometimes referred to as paratuberculosis.

____

10. _____ are one of the most destructive external parasites in dairy herds, as they can cause permanent damage to teats.

____

11. List ways producers can prevent lameness in dairy herds.

12. Outside of vital signs, what other health considerations might a producer check in dairy cattle?

13. _____ is an inflammation and infection of the uterus that can be caused by a retained placenta or an inability to combat bacteria entering the uterus during calving.

Key Terms

Carefully study the chapter and then use the word bank to match the term to the correct definition.

A. biometrics	D. quarters	G. clinical	J. colostrum
B. lameness	E. mastitis	H. ketosis	
C. sole	F. subclinical	I. keratin	

1. _____ The bottom of the hoof
2. _____ Information about what the cow did during the day, including the amount of time spent walking, resting, or eating, the amount of food and water consumed, and the amount of milk produced
3. _____ Infection lasting less than two months
4. _____ Any impediment to a cow's ability to move freely in her environment
5. _____ Four sections of the udder that each end in a teat
6. _____ Infection lasting long periods of time or even for the remainder of life
7. _____ Provides immunity and extra nutrition to the calf in its first few days of life
8. _____ An infection in the udder that occurs when pathogens enter the teat through the streak canal and work their way up into the milk-producing alveoli
9. _____ A semi-hard material that is constantly growing and, like your fingernails, may be very thin or thick
10. _____ A metabolic disorder characterized by a lack of nutrients and a negative energy balance which can lead to severe weakness and death

Critical Thinking

Using the information learned in this chapter, read the question and provide a detailed answer.

1. Dairy cattle live longer lives than many production animals. What challenges does this pose in managing and caring for a dairy herd?

Name ______________________ Date ____________ Period ________

Activity: Dairy Cattle Management and Care

Evaluating Body Condition of Dairy Cattle

Required Materials

- Laboratory notebook
- Pen or pencil
- Farm or laboratory outfit
- Elanco™ Body Condition Score Guide

Activity Specification

The body condition score (BCS) of dairy cattle provides farmers and animal scientists with a long-term perspective into the nutritional well-being of dairy cattle and of the events that have led cattle to look the way they do. BCS can be used by farmers to make management decisions related to nutrition and breeding.

Use the following general steps to give dairy cattle a body condition score. Consult your text and the Elanco™ Body Condition Score Guide if you need additional reference points.

1. View cattle from the side. Check the zone between the hooks, pins, and hooks. A very flat "v" means the BCS is 3.0 or greater.
2. Identify the pelvic hooks of cattle. If angular, the BCS is less than 3.0. If smooth, the BCS is 3.0.
3. If hooks are angular, but pins are not too prominent, BCS is 2.75.
4. If hooks are angular and pins are prominent, BCS is 2.5. Ribs and vertebrae might begin to be visible.
5. If the animal has less fat than a BCS of 2.5 and the ribs are largely visible, the BCS is 2.25 or lower.
6. If the zone between the hooks, pins, and hooks looks like a flattened "U" and pins are beginning to disappear, the BCS is at least 3.25.
7. If the sacral ligament begins to disappear, the BCS is at least 3.5.
8. More fat and the disappearance of the sacral ligament indicate a BCS greater than 4.

Evaluate three dairy heifers or cows, based on live animals or pictures. Discuss your reasoning for giving each animal a score.

Animal #1

1. Determined BCS: ______________________
2. Explanation:

____

____

Animal #2

1. Determined BCS: ______________________
2. Explanation:

____

____

Animal #3

1. Determined BCS: ____________________
2. Explanation:

Reflective Questions

Answer the following questions relating to the activity.

1. Analyze your classmates' rankings of cattle body condition scores. Did everyone come to the same conclusions? What differences did you see, if any?
2. After seeing how your instructor ranks the animals, evaluate your performance. In what areas did you perform closest to them? In what areas did you have trouble?
3. Create a hypothesis about what life cycle stage each of the cows is in, according to her apparent BCS.

Name ______________________ Date ____________ Period ________

CHAPTER 36 Dairy Cattle Selection and Evaluation

Carefully study the chapter and answer the following questions.

1. What program encourages dairy producers to test their herds and keep records?

2. List the major factors considered in dairy cattle conformation.

3. The _____ should be prominent and evenly set on both sides of the body.

4. To evaluate dairy strength, the overall size and shape of the _____, _____, and _____ are major considerations.

5. What characteristics indicate overall udder health and productivity?

6. The _____ and _____ should be most nearly level when viewed in profile.

7. PDCA is an acronym for _____.

8. List the factors used to determine the dairy strength of an individual animal.

9. What form is used to evaluate conformational characteristics for dairy cattle?

10. A dairy cow's _____ should be nearly level and set wide.

11. What is the single most important factor to examine when judging a dairy cow?

Key Terms

Carefully study the chapter and then use the word bank to match the term to the correct definition.

A. dairy herd record
B. dairy strength
C. frame
D. fore udder attachment
E. lateral suspensory ligaments
F. mammary vein
G. median suspensory ligament
H. milk wells
I. thurl
J. udder cleft

1. _____ The data kept for a milking herd related to individual performance
2. _____ Supports the udder and runs through the center, holding its weight
3. _____ Term used to describe how much the conformation of a cow is suited to milk production
4. _____ Place where veins pass through the wall of the udder
5. _____ The skeletal structure of dairy cows
6. _____ Runs perpendicular to the ground and helps support each quarter of the udder
7. _____ Place on the rump where the leg attaches to the pelvis
8. _____ Large blood vessels which provide nutrients and water to the udder for milk production
9. _____ An examination of how well supported the udder is and the strength of the suspensory ligament
10. _____ The front section of the udder closest to the cow's belly

Critical Thinking

Using the information learned in this chapter, read the question and provide a detailed answer.

1. Why are frame size and rear leg conformation so important to selecting and evaluating dairy cattle?

____

____

____

____

____

Name ______________________ Date ______________ Period __________

Activity: Dairy Cattle Selection and Evaluation

Dairy Cattle Showmanship

Required Materials

- Laboratory notebook
- Pen or pencil
- Farm or laboratory outfit
- Halter
- Show stick

Activity Specification

Dairy cattle showmanship is an activity performed by members of agriculture organizations such as FFA and 4-H. Today, your instructor will demonstrate the basics of showing dairy cattle. You will take notes below before being tasked with showing a heifer or cow yourself!

NOTES:

Reflective Questions

Answer the following questions relating to the activity.

1. Analyze the behavior of dairy cattle as they are led across the arena. What are three factors that seem to distract their performance?

2. Evaluate the body posture of the students showing the cattle. What attributes of their posture stand out to you?

3. Create a list of five tips that new handlers can employ when showing their dairy cattle.

Name ______________________ Date ____________ Period ________

CHAPTER 37 Classification and Breeds of Dairy Cattle

Carefully study the chapter and answer the following questions.

1. _____ heifers mature at a slightly younger age than other breeds, allowing young cows to enter the milking herd sooner.

____

2. This breed originated in the Channel Islands just off the coast of France.

____

3. _____ is one of the most common dairy breeds in Australia.

____

4. This breed was originally imported into the United States in 1869, with 150 animals forming the foundation for the 55,000–85,000 head here today.

____

5. _____ are a cross between *Bos indicus* Gyr cattle from India and Holsteins.

____

6. Around _____% of all dairy cattle in the United States are black and white _____.

____

____

7. _____ are a deep red breed known to be efficient grazers; well suited to cold, wet climates; with superior udder characteristics.

____

8. In 1964, a group of Shorthorn breeders founded the _____, which provided a platform for creating and promoting crossbred cattle that were red and white.

____

9. This breed originated in the Netherlands around 2000 years ago.

____

10. _____ cattle have the highest milkfat content in their milk.

____

11. More than _____% of all American dairy cattle come from just _____ breeds.

____

____

12. List the seven major US dairy cattle breeds.

_______________________________ _______________________________

_______________________________ _______________________________

_______________________________ _______________________________

____

13. This breed averages 23,000 pounds of production per lactation cycle with an average milkfat of 3.8%.

____

14. The average Holstein produces _____ pounds of milk per production cycle.

____

15. This breed is sought after as ideal for cheesemaking.

____

Key Terms

Carefully study the chapter and then use the word bank to match the term to the correct definition.

A. Holstein	C. Red and White	E. Ayrshire	G. Milking Shorthorn
B. Jersey	D. Brown Swiss	F. Guernsey	H. Red Sindhi

1. _____ This breed is known for their prized "golden" milk and easy management qualities
2. _____ This breed is the result of a recessive gene in the coat patterns of other dairy breeds
3. _____ This breed is the second most common in the United States
4. _____ This breed is distinguished by a black muzzle encircled by a white ring
5. _____ One of the oldest breeds of dairy cattle that originated as a dual-purpose breed
6. _____ Originating from Pakistan, this breed is widely used in crossbreeding programs in the Indian dairy cattle industry
7. _____ This breed is second in size to Holsteins
8. _____ Originated as a cross between Batavian black cattle and Friesian white cattle

Critical Thinking

Using the information learned in this chapter, read the question and provide a detailed answer.

1. Why is the US dairy industry dominated by just seven breeds of cattle? Why is one so much more popular and widely used than all the others?

____

____

____

____

Name ______________________ Date ____________ Period ________

Activity: Dairy Cattle Classification and Breeds

Presenting Dairy Cattle Breeds

Required Materials

- Computer with internet access
- Gmail account

Activity Specification

Today you will create a PowerPoint® or Google Slides presentation on dairy cattle, showcasing at least six different breeds.

Your PowerPoint should:

- Include 10 slides in total
    - Include a title page and a reference page
- Introduce the scientific name of each animal and any appropriate scientific terms (age/sex)
- Define the average ages of puberty and breeding and length of gestation
- Describe six different dairy cattle breeds
    - Explain breed details and their purposes

At the conclusion, be prepared to give a presentation to the class, covering what you learned.

Reflective Questions

Answer the following questions relating to the activity.

1. Analyze your classmates' presentations. Is there a common theme regarding the most common breeds of dairy cattle used?

____

____

____

2. Evaluate your own presentation and indicate at least two aspects where you could have improved your communication of livestock breeds.

____

____

____

3. Create a picture of what your ideal dairy cow breed would look like and describe three important traits that make this dairy cow breed ideal.

Name ______________________ Date ____________ Period ________

CHAPTER 38 Equine Industry Overview

Carefully study the chapter and answer the following questions.

1. The first three distinct types of horses developed were:

   A. ______________________, used for war

   B. ______________________, used for traveling

   C. ______________________, used for speed

2. _____ were the first equine species to be domesticated around _____ BCE.

   ____

   ____

3. Major sectors of the equine industry include:

   ______________ ______________

   ______________ ______________

   ____

4. The _____ are widely considered the first culture to master horsemanship.

   ____

5. The three largest equine breed associations in the United States by overall number of registered animals are:

   ______________ ______________

   ____

6. The equine industry provides more than $_____ in direct economic impact to the United States each year.

   ____

7. Describe the differences between English and Western riding disciplines.

   ____

   ____

8. _____ are the bucking animals that are ridden during rodeo events.

   ____

9. The _____ is a national organization that represents the interests of all segments of the equine industry in the United States.

   ____

10. Describe animals in the family *Equidae*.

    ____

    ____

    ____

11. Records of cart horses exist from as early as _____ BCE.

____

12. The top five states by number of horses are:

_______________________ _______________________

_______________________ _______________________

____

Key Terms

Carefully study the chapter and then use the word bank to match the term to the correct definition.

A. broodmare	D. filly	G. hinny	J. hippotherapy
B. colt	E. foal	H. john	
C. stockhorse	F. gelding	I. mare	

1. _____ Mature male mule or hinny
2. _____ Equine-assisted therapy
3. _____ Female equine two years or younger
4. _____ Offspring of a female ass or donkey and a male horse
5. _____ Mature female horse
6. _____ Young equine of either sex
7. _____ Castrated male equine
8. _____ Mature female horse kept for breeding
9. _____ Male equine two years or younger
10. _____ Horse that has been selected for ranch working traits

Critical Thinking

Using the information learned in this chapter, read the question and provide a detailed answer.

1. The chapter notes that many horse owners keep their animal(s) for more than one purpose. Why do you think this might be?

____

____

____

____

____

____

____

Name ______________________________ Date ______________ Period __________

Activity: Equine Industry Overview

Outfitting a Horse Properly

Required Materials

- Laboratory notebook
- Pen or pencil
- Farm or laboratory outfit with gloves
- Horse or horse-like prop
- Saddle
- Bridle
- Girth
- Breastplate
- Fly mask

Activity Specification

Today you will get hands-on experience with equipping a horse for riding after being instructed by your teacher. In groups of four or five students, you will be required to demonstrate mastery of calmly applying each of the provided materials on a class horse or prop.

Your instructor will demonstrate how to safely:

- Place and fasten a saddle and girth.
- Put on a bridle.
- Put on a breastplate.
- Put on a fly mask.
- Take notes as your instructor completes the demonstration.

Reflective Questions

Answer the following questions relating to the activity.

1. Analyze the structure of a saddle. What aspect of its placing is the most complicated? How can you ensure that this placement is done in a calm and safe manner?

____

____

____

2. Evaluate your timeliness in setting up a horse or prop. If you were not in groups and had to set up independently, what would you do differently to ensure safety?

____

____

____

3. Create a drawing below where you draw a horse with the tack you learned to put on today. Clearly label each piece in the drawing.

Name ______________________ Date ____________ Period ________

CHAPTER 39 Equine Production Systems

Carefully study the chapter and answer the following questions.

1. Horses often exhibit _____ behaviors, as seen when horses scratch each other's withers to bond.

____

2. _____ are small enclosures that are bolted firmly to the ground to restrain a horse.

____

3. Why do equine producers strive to time foaling to be as close to January 1 as possible?

____

____

4. _____ is the term used to describe a horse who can be haltered, led, and tied safely.

____

5. A _____ is a small enclosure used for housing or exercising horses.

____

6. Equine gestation periods are based on the _____ and _____ of the animal.

____

____

7. _____ are commonly referred to as backyard or hobby breeders.

____

8. _____, _____, and _____ have gestation periods between 345 and 385 days.

____

____

____

9. Most horses receive some training shortly after birth, called _____, to ensure that they are accustomed to human interaction and will be safe to handle as they grow.

____

10. Almost _____% of horse sales involve social media.

____

11. A mare will typically carry a foal between _____ and _____ days.

____

____

12. _____ is the equipment needed to handle, ride, or harness a horse.

____

13. List the three main categories of large-scale equine production.

______________________________ ______________________________

____

Key Terms

Carefully study the chapter and then use the word bank to match the term to the correct definition.

A. bit	D. broke	G. finished	J. stud
B. box stall	E. cross-tie	H. groundwork	
C. bridle	F. green	I. hitching post	

1. _____ An enclosed area within a barn where an animal is kept
2. _____ A horse who has completed advanced training in a specific discipline
3. _____ Training practices without a rider on top of the horse
4. _____ A horse that has not had much training and is new to being ridden or driven
5. _____ A horizontal pipe set to hit at chest height on most horses
6. _____ The place where a stallion is kept for breeding
7. _____ Mouthpiece used on some bridles to cue a horse while riding
8. _____ A horse that has completed basic training under saddle
9. _____ An option for tying horses that uses two separate lead ropes or tethers that attach to the sides of the horse's halter
10. _____ Used to give guidance to a horse's head while riding

Critical Thinking

Using the information learned in this chapter, read the question and provide a detailed answer.

1. What considerations do owners and producers need to make in designing housing for equine animals?

____

____

____

____

Name ______________________ Date ______________ Period __________

Activity: Equine Production Systems

Leading a Horse

Required Materials

- Laboratory notebook
- Pen or pencil
- Farm or laboratory outfit with gloves
- Horse
- Saddle
- Bridle with lead
- Girth
- Breastplate

Activity Specification

Now that you are familiar with the tack used to saddle a horse, you will once again demonstrate your ability to do this. Once you have completed this first task, your instructor will give you a course to lead your horse through.

Draw the path that your instructor wants you to guide your horse through in the area provided below:

Goodheart-Willcox Publisher

Reflective Questions

Answer the following questions relating to the activity.

1. Analyze the eyes and ears of a horse as the instructor or your classmates leads it. What correlation (if any) do you see between how the horse's ears and eyes are moving and what it is doing?

2. Evaluate the movement of a horse and the different types of strides/walking patterns it takes as it is being led by the instructor. What patterns do you notice in its steps?

3. Draw an image below to indicate areas of danger when walking around horses. Use other knowledge of livestock to inform your drawing.

Name ______________________ Date ____________ Period ________

CHAPTER 40 Equine Management and Care

Carefully study the chapter and answer the following questions.

1. _____ is caused by a virus that is transmitted through direct contact between horses and animals or environments which carry the virus. Its symptoms include fever, persistent cough, nasal discharge, and loss of appetite.

____

2. _____ is a virus commonly found in birds, then passed to other animals by mosquitoes.

____

3. What factors indicate that a horse should be vaccinated frequently?

______________________ ______________________

____

4. Some horse owners and many mule owners choose to completely remove the mane with clippers, a process called _____.

____

5. Horses have remarkably sensitive _____, which trigger a fight-or-flight response.

____

6. At what age are equine skeletons considered mature?

____

7. List the three stages of equine parturition.

______________________ ______________________

____

8. _____ is sometimes referred to as "sleeping sickness" and comes in several forms, all of which attack the brain and central nervous system.

____

9. Describe why grooming equines is important.

____

____

10. Briefly describe what is unique about the digestive system of a horse compared to other domesticated animals, and what that means for its dietary requirements.

____

____

11. _____ is an inflammation of the soft tissue in the hooves that can cause irreparable damage and permanent lameness.

____

12. The _____ is a pouch-like area found at the beginning of the large intestine, where digestive microbes are housed to break down _____.

13. _____ has no vaccination or effective treatment, and once the horse is infected it will be a lifelong carrier of the disease.

14. Up to _____ equine vaccines may be combined in a single dose.

Key Terms

Carefully study the chapter and then use the word bank to match the term to the correct definition.

A. bars	D. farrier	G. frog	J. white line
B. colic	E. floating	H. Galvayne's groove	
C. cups	F. freeze branding	I. hand breeding	

1. _____ V-shaped depressions in the sole of the hoof
2. _____ An indentation on the rear incisor
3. _____ The process of evening the surface of equine teeth
4. _____ The place where the outer layer and inner layer of the hoof attach to one another
5. _____ Involves the use of chemicals to chill a branding iron before placing it on a horse's skin
6. _____ A professional who specializes in the care of hooves
7. _____ Any obstruction in the digestive tract of a horse that prevents the passage of food, gas, or both
8. _____ Indentations in the incisors
9. _____ A triangular area in the middle of the sole
10. _____ Bringing the mare and stallion together to breed every other day if both are receptive to breeding

Critical Thinking

Using the information learned in this chapter, read each question and provide a detailed answer.

1. Select an illness covered in this chapter. Describe the symptoms caretakers should look for in diagnosing the condition, and how it is transmitted.

2. What components of a herd management program are vital to prevent equines from contracting this illness?

Name ______________________ Date ____________ Period ________

Activity: Equine Management and Care

Hoof Care

Required Materials

- Laboratory notebook
- Pen or pencil
- Farm or laboratory outfit with gloves
- Horse
- Hoof tester
- Rasp
- Farrier knife

Activity Specification

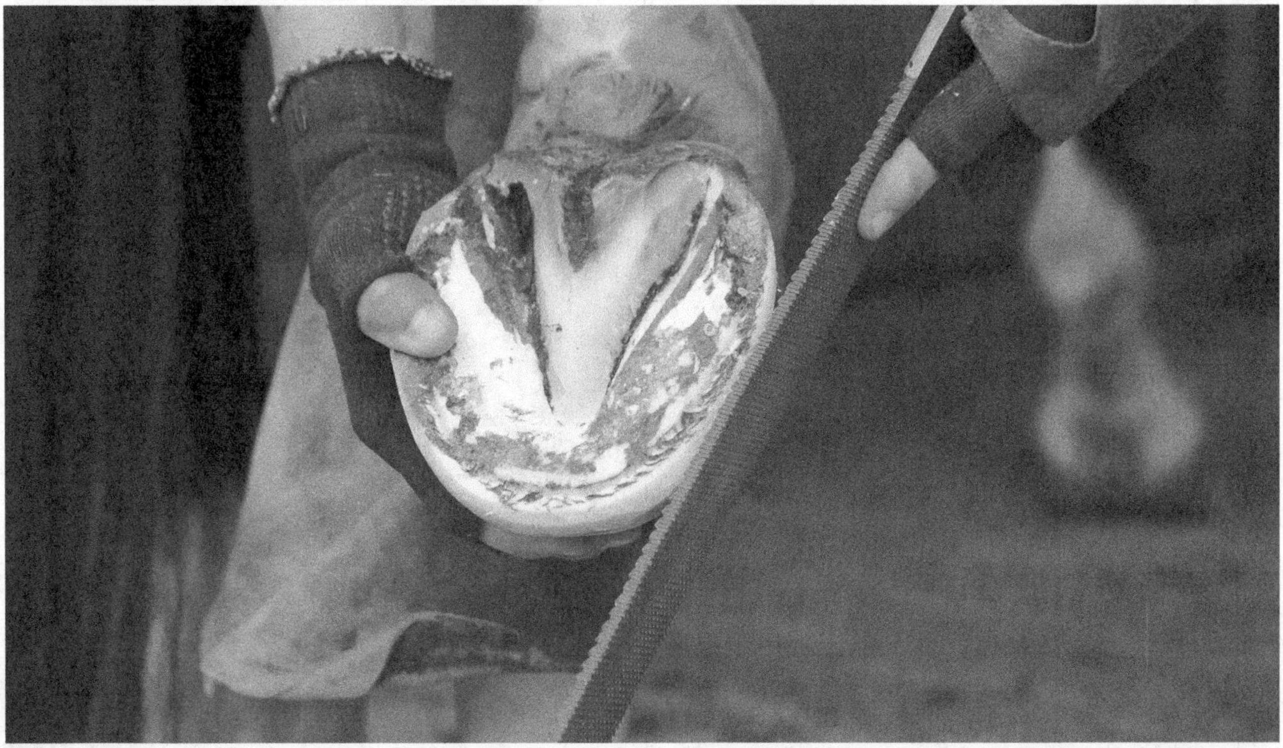

anjajuli/Shutterstock.com

Farriers play an important role in maintaining equine hoof health. Farriers assist with evaluating, trimming, shoeing, and maintaining horses' hooves. Today you will learn how to clean a horse's hooves. Your instructor will demonstrate the proper method of grasping a horse's leg between your knees and the simple tools that can be used to inspect and clean residue from a hoof. Take notes about what you observe.

NOTES: ____

Reflective Questions

Answer the following questions relating to the activity.

1. Analyze particulate build-up found in a horse's hoof. Describe the appearance of this build-up and any other areas of damaged integrity.

2. Evaluate the comfort of a horse while its leg is being inspected. What is necessary during this process to ensure safety?

3. Create a drawing showing the hoof that you inspected and cleaned, before and after.

Before:	After:

Name ______________________ Date ____________ Period ________

CHAPTER 41 Equine Selection and Evaluation

Carefully study the chapter and answer the following questions.

1. _____ is the first thing to examine when evaluating equines.

____

2. _____ is the term used to describe the cream, ivory, or silver mane and tail coloring that occasionally presents with chestnut, sorrel, or chocolate coat colors.

____

3. The five most important places to examine muscling on horses are:

______________ ______________

______________ ______________

____

4. Excessive _____ markings can exclude horses from registration in many breed associations.

____

5. What is a gait, and what are the two types of gaits? What differences exist between the two?

____

____

____

____

6. A _____ is a small grouping of white hairs on the forehead.

____

7. A _____ is an equine measurement equaling four inches.

____

8. Height is measured at the horse's _____, where the _____ meets the back.

____

____

9. A(n) _____ is a cosmetic condition that does not interfere with performance, while a(n) _____ is a condition that interferes with the horse's ability to perform its required tasks.

____

____

10. Some breeds recognize a _____, which is a white mark that extends halfway between the hoof and the pastern joint.

____

11. Points can be found on the horse in these places:

_______________ _______________

_______________ _______________

____

12. Good balance is shown when equines are evenly proportioned across three sections:

_______________ _______________

____

13. Most experts identify five natural gaits. What are they?

_______________ _______________

_______________ _______________

____

Key Terms

Carefully study the chapter and then use the word bank to match the term to the correct definition.

A. overo	C. buckskin	E. chimera	G. dappling
B. tobiano	D. chestnut	F. coronet	

1. _____ Marking pattern that does not feature white crossing the animal's back between its withers and tail, but does feature at least one dark leg; irregular or splotchy white markings; a distinct face pattern; and a solid color tail
2. _____ A thin white band above the hoof
3. _____ The presence of lighter-colored circles on a darker base coat
4. _____ An animal that possesses two different genetic makeups
5. _____ Marking pattern that usually features dark colors covering one or both flanks; four white legs below the knees; regular, distinct spots that extend over the neck and chest; and often a bi-colored tail
6. _____ A color ranging from light red-brown to dark red or liver-colored
7. _____ Golden- to light-tan body color with black mane and tail

Critical Thinking

Using the information learned in this chapter, read the question and provide a detailed answer.

1. Equine selection and evaluation are handled differently than other domesticated species. What are the main similarities and differences that you note?

Activity: Equine Selection and Evaluation

Evaluating Gait

Required Materials

- Laboratory notebook
- Pen or pencil

Activity Specification

Today you will evaluate the four primary forms of equine movement. The way a horse moves is called its *gait.* There are four primary forms of gaits:

- Walk: Each foot hits the ground separately. This is the slowest of gaits, with an even rhythm.
- Trot: A trot is characterized as a two-beat gait. Generally, a front leg and a diagonal back leg move together in opposing sequence. This gait is at a moderate speed (faster than a walk) and classically causes the horse to move up and down.
- Canter: The canter is faster than a trot and is a three-beat gait.
- Gallop: The gallop is the fastest of the four primary gait forms. In this gait, the hind legs propel the horse forward while the front legs catch its forward motion and stabilize it before the hind legs propel it forward again.

The descriptions above described the foot movement for the horse in all gaits but one. Your job today is to describe how a horse's legs move when it canters.

1. Draw a horse cantering below.

2. Describe the stepping sequence of a canter.

Reflective Questions

Answer the following questions relating to the activity.

1. What was difficult about describing the canter compared to other gaits? Why?

2. When do horses move in each of the different gaits?

3. How does a horse's movement contribute to evaluating its conformation? What conformational factors are especially important for movement?

Name ______________________ Date ____________ Period ________

CHAPTER 42 Classification and Breeds of Equines

Carefully study the chapter and answer the following questions.

1. The _____ descended from the feral horses of the American West and must have come directly from the range or trace its parentage to the range to qualify for registration.

   ____

2. The _____ breed originated from a combination of donkey breeds imported into the United States.

   ____

3. *Equus caballus* species are split into four different categories:

   ______________ ______________

   ______________ ______________

4. The _____ breed is heavy boned, muscular, has a thick neck and strong topline, and is primarily used for youth riders.

   ____

5. The _____ breed originated in the colonial United States with superior riding characteristics.

   ____

6. The _____ breed originated in Scotland, where it was used as a farm horse and is well-known for the excess hair, or feathering, on its lower legs.

   ____

7. Registered animals are sometimes referred to as _____ animals because they have _____ from the breed association.

   ____

   ____

8. The _____ has the largest breed registry in the United States and is the most popular breed in the nation.

   ____

9. The _____ is characterized by its gray color and small stature.

   ____

10. _____ include long-eared equines like donkeys and burros.

    ____

11. _____ are horses of all breeds and sizes.

    ____

12. The _____ breed descended from the horses of the Nez Perce Native American tribe and its coloring often includes a spotted coat pattern.

    ____

13. The _____ breed originated in Argentina in the fifteenth century.

____

14. The _____ is a color breed for horses who combine stockhorse conformation with pinto coat patterning.

____

15. The _____ breed are typically black or gray and originated in Western France, where they were used for battle, farm work, and carriage pulling.

____

Key Terms

Carefully study the chapter and then use the word bank to match the term to the correct definition.

A. burro	D. hotblood	G. grade horse
B. coldblood	E. light horse	H. warmblood
C. draft horse	F. pony	I. miniature horse

1. _____ A term used for draft horses
2. _____ Horses that were bred to be used under saddle and typically stand between 14 and 16 hands
3. _____ Horses that are not registered with any breed association
4. _____ The largest type of horse, typically standing 16 to 18 hands
5. _____ A term used for more spirited light breeds
6. _____ The name for a standard donkey that originated in North America
7. _____ Used to define a subcategory of light horses whose lineage includes draft horse breeds
8. _____ The smallest type of horse; cannot stand taller than 34 inches at the withers
9. _____ Classification given to horses less than 14.2 hands tall with stocky bodies and heavy bones

Critical Thinking

Using the information learned in this chapter, read the question and provide a detailed answer.

1. Equines are an incredibly versatile species. While equine meat is not consumed in the United States, how many ways have equines helped to feed the US population?

____

____

____

____

____

____

Name ______________________________ Date ______________ Period __________

Activity: Classification and Breeds of Equines

Showcasing Equine Breeds

Required Materials

- Computer with internet access
- Gmail account

Activity Specification

Today you will create a PowerPoint® or Google Slides™ presentation about horses, while showcasing at least six different breeds of horses.

Your presentation should:

- Consist of 10 slides total
    - Include a title page and a reference page
- Introduce the scientific name of each animal and any other relevant terms
- Explain the average age of puberty and breeding and length of gestation
- Describe six different equine breeds
    - Describe body type, classification, coloring, temperament, size, historical information, and the uses or purposes of the breed

Give a five-minute presentation to the class where you present your slideshow and describe the breeds that you have chosen to discuss.

Reflective Questions

Answer the following questions related to the activity.

1. Analyze your classmates' presentations. Are there any common themes regarding the breeds that were presented?

____

____

____

____

2. Evaluate your own presentation and indicate at least two aspects where you could have improved your communication.

____

____

____

____

3. Create a picture of what your ideal equine breed would look like. Describe three traits that would make this horse breed unique.

Name ______________________________ Date ______________ Period __________

CHAPTER 43 Poultry Industry Overview

Carefully study the chapter and answer the following questions.

1. Of the $46.1 billion that the poultry industry produced in 2021, _____% of that value came from broiler production and sales ($_____).

2. Why must US consumers refrigerate their eggs, while consumers in other countries do not?

3. The _____ is the oldest livestock association in the United States, originally founded in _____.

4. _____ was the first person to commercially produce poultry after a miscommunication had _____ chicks delivered to her home.

5. One of the most expensive and popular cuts of meat from chickens is the _____, which used to be deemed worthless and thrown out.

6. The unfertilized eggs we eat are also called _____ or _____eggs.

7. Chickens are believed to be the living descendants of _____.

8. _____ established the modern, vertically integrated poultry industry in _____.

9. _____ products are those that have been marinated, breaded, pre-cooked, or altered to be sold as a more finished product.

10. It takes just over _____ for an egg to develop within a hen's reproductive tract.

11. Why do chickens display cannibalistic tendencies?

12. _____ is known as the "all feather" organization and was founded in _____.

13. The _____ is the breed from which all domesticated chickens descended.

14. US table eggs are gathered, washed, sanitized, packed, and shipped to the grocery store within _____ of laying.

Key Terms

Carefully study the chapter and then use the word bank to match the term to the correct definition.

A. breeder	D. hen	G. pullet	J. vertical integration
B. cockerel	E. poult	H. tom	
C. free-range	F. poultry	I. cage-free	

1. _____ Mature male turkey
2. _____ Chicken that is kept as genetic stock for future broilers, table egg layers, or other purposes
3. _____ Young female chicken less than one year of age
4. _____ Descriptor of table egg layer flocks that are raised in poultry houses with cages containing multiple birds and the ability to maneuver across the house and spend time on the ground
5. _____ Mature female chicken
6. _____ Multiple phases of a business or industry conducted by a single company
7. _____ Young turkey less than one year of age
8. _____ Poultry production that must include free access to the outdoors, with exposure to fresh air, grass, and insects in a more natural environment
9. _____ Young male chicken less than one year of age
10. _____ Birds that have been domesticated for human use or consumption

Critical Thinking

Using the information learned in this chapter, read the question and provide a detailed answer.

1. Poultry products often feature many terms and labels to entice consumers to purchase. Do you agree with the use of labels like "No Added Hormones" or "Non-GMO" on poultry products? Why or why not? Define these terms in your answer.

Name ______________________ Date ______________ Period __________

Activity: Poultry Industry Overview

Determining Egg Quality

Required Materials

- Laboratory notebook
- Pen or pencil
- Farm or laboratory outfit with gloves
- Basket of 15 eggs
- Air cell gauge
- Scale
- Egg candler
- Tripod micrometer

Activity Specification

Today your instructor will demonstrate several procedures for determining the quality of eggs. When you are done, what can you do with the eggs?

1. Randomly number eggs from 1 to 15.
2. Separate the eggs based on size. Make notes about each egg in your log.
3. Inspect each egg for shell inconsistencies; separate eggs with questionable shells from the group. Make notes about these eggs in your log.
4. Use an egg candler to view the inside of each egg and determine the yolk integrity. Make notes in your log.
5. Using an air cell gauge, determine the quality of the egg. Make notes in your log.
6. Weigh each egg. Make notes in your log.
7. Crack eggs open and use a tripod micrometer to measure the albumen thickness. Make notes about this measurement in your log.

Egg #	Size (s-small, m-medium, l-large, el-extra large, j-jumbo)	Smooth shell (y-yes, n-no)	Yolk integrity	Air cell quality (AA, A, B)	Weight of eggs (grams)	Height of albumen (mm)
1						
2						
3						
4						
5						
6						
7						
8						
9						
10						
11						
12						
13						
14						
15						

Reflective Questions

Answer the following questions relating to the activity.

1. Analyze the data you have obtained. What are two patterns that you recognize across the dataset?

2. Evaluate the size of eggs as it relates to the other measurements. Do you see any correlations? Is bigger always better? Why or why not?

3. Create drawings that illustrate what the inside of a small AA egg and a jumbo B egg would look like.

Small AA	Jumbo B

Name ______________________________ Date ______________ Period __________

CHAPTER 44 Poultry Production Systems

Carefully study the chapter and answer the following questions.

1. Feed makes up about _____% of the cost of growing broilers.

____

2. For the first week of life, chicks are _____, meaning they cannot regulate their own body temperature and depend on supplemental heat.

____

3. Describe the safest way to hold a chicken.

____

4. The ideal temperature for chick embryo development is _____° Fahrenheit.

____

5. Name six normal activities and behaviors a farmer would observe daily in a broiler breeder or cage-free laying hen house.

______________________ ______________________

______________________ ______________________

______________________ ______________________

6. Embryos will communicate with each other by _____ to either accelerate or slow down development to synchronize the hatch.

____

7. Hens can store sperm for up to _____ following mating.

____

8. What are the two main phases of management for broiler breeders? Include the terminology for the birds and age ranges for each phase.

____

____

9. The mating process of chickens is also known as a "_____."

____

10. It is now thought that chickens produce at least ten distinct forms of vocalization, including the following calls:

______________________ ______________________

______________________ ______________________

______________________ ______________________

______________________ ______________________

______________________ ______________________

11. Modern broilers originated from a cross of what two breeds?

____

12. Describe the difference between primary and further processing.

____

____

13. Fertile eggs are collected _____ times per week from broiler breeder farms and are held in dedicated egg coolers at _____ ° Fahrenheit.

____

____

14. House floor temperatures should be around _____ ° Fahrenheit on the day that broiler chicks arrive from the hatchery to optimize feed conversion and growth.

____

15. What is FCR and how is it calculated?

____

____

Key Terms

Carefully study the chapter and then use the word bank to match the term to the correct definition.

A. aviary system	D. floor system	G. preening	J. convertible colony cage system
B. broiler breeder	E. *ad libitum*	H. integrator	
C. brooding	F. photostimulation	I. pecking order	

1. _____ System in which hens have external perches that allow for movement in the house when the doors are open
2. _____ System that allows for increased horizontal movement but less vertical movement, similar to broiler breeder housing
3. _____ Spreading oil across a bird's feathers to keep them in good condition
4. _____ Process that occurs when pullets and cockerels are provided more than 12 hours of light to stimulate sexual maturation
5. _____ The companies involved in raising, processing, and marketing broilers
6. _____ Social organization within a flock
7. _____ System in which hens are provided with a series of multi-tiered platforms with perches
8. _____ Feeding without limits
9. _____ The first week of raising broilers, when supplemental heat must be provided
10. _____ Chickens used to produce fertile eggs to ensure a steady supply of broilers for meat production

Name ____________________

Critical Thinking

Using the information learned in this chapter, read the question and provide a detailed answer.

1. Describe, in your own words, the pros and cons of cage vs. cage-free systems used in shell egg production. Include details about different types of systems.

Name ______________________ Date ______________ Period ________

Activity: Poultry Production Systems

Meat Cuts from a Broiler Carcass

Required Materials

- Laboratory notebook
- Pen or pencil
- Farm or laboratory outfit with gloves
- Trash bucket
- 1 whole chicken
- Mesh gloves
- Knife assortment

Activity Specification

Today, your instructor will demonstrate how to take a whole raw chicken carcass purchased from the store and cut it into retail cuts.

Utawi Studio/Shutterstock.com

After watching the demonstration, in groups of three, work together to produce the following cuts from your broiler:

- 2 chicken breasts
- 2 chicken tenderloins
- 2 drumsticks
- 2 thighs
- 2 drumettes
- 2 wings

Reflective Questions

Answer the following questions relating to the activity.

1. Analyze the carcass after you have trimmed away the fabricated cuts. How much usable material remains?

 ____

2. Evaluate the cleanliness and efficiency of your cuts of meat. What are two ways your butchering skills could improve?

 ____

 ____

3. List four meals that can be created with the cuts of meat that you have trimmed today in lab.

 ____________________ ____________________

 ____________________ ____________________

Name ______________________ Date ____________ Period ________

CHAPTER 45 Poultry Management and Care

Carefully study the chapter and answer the following questions.

1. _____ is a disease spread by the *Eimeria* oocysts found in infected bird droppings.

____

2. The five yellow, yolk-filled follicles in the reproductive system of a hen are known as the _____.

____

3. The _____ is a long, narrow slit that divides the hard palate and connects the trachea to the nasal passages.

____

4. The _____ publishes detailed nutritional requirements for different poultry species, types of producers, and ages of birds.

____

5. The _____ or _____ is a thin protein coat applied around the porous eggshell, sealing it and protecting the egg from water loss and microbial invasion.

____

____

6. _____ has the potential to be a catastrophic disease in commercial poultry, with mortality rates close to 100%.

____

7. The small intestine, the primary site for nutrient absorption, is divided into three sections:

A. ______________________ C. ______________________

B. ______________________

8. What is the importance of preening to poultry?

____

____

____

9. Bird skeletons contain a pair of bones in the shoulder area called _____ that prevent the wing muscles from compressing the chest during a powerful downstroke.

____

10. _____ is a chronic respiratory disease caused by the bacterium *Mycoplasma gallisepticum*.

____

11. Why are mineral supplements important to raising poultry?

____

____

____

12. The hormone _____ stimulates the small ovarian follicles to grow and secrete sex hormones, including _____.

13. The principal muscles for flight are the _____, marketed as chicken breast, and the _____, marketed as the chicken tender.

14. The _____ is where the mechanical action of mixing and grinding ingested feed occurs.

15. _____ is a bacterial disease of the intestinal tract that presents with a rapid onset of clinical signs such as ruffled feathers and bloody diarrhea and is quickly followed by an increase in mortality.

Key Terms

Carefully study the chapter and then use the word bank to match the term to the correct definition.

A. air sac system
B. germinal disc
C. clutch
D. myoglobin
E. cloaca
F. proventriculus
G. *in ovo* vaccination
H. phase feeding
I. hatchability
J. physiological zero

1. _____ A vaccine that is injected into a developing embryo, usually at day 18 of incubation during transfer from setter to hatcher
2. _____ The temperature at which development pauses and the embryo becomes dormant
3. _____ Where gastric juices are produced and secreted
4. _____ A chamber shared across the digestive, excretory, and reproductive systems that opens externally at the vent
5. _____ Its increased presence gives dark meat its color
6. _____ The ratio of chicks hatched to the number of eggs set
7. _____ A system of making specific feeds to meet the changing nutritional requirements of the different stages of production
8. _____ This respiratory system addition sets birds apart from other mammals
9. _____ Appears as the white spot on the surface of the egg yolk and contains the female pronucleus
10. _____ The pattern in which a hen lays eggs

Critical Thinking

Using the information learned in this chapter, read the question and provide a detailed answer.

1. Consider the labels seen on egg cartons at the grocery store. After learning more about egg production in the chapter, do you notice anything confusing or even misleading about the way that eggs are marketed?

Name ______________________ Date ____________ Period ________

Activity: Poultry Management and Care

Poultry Handling

Required Materials

- Laboratory notebook
- Pen or pencil
- Farm or laboratory outfit with long gloves

Activity Specification

Poultry handling is an essential technique that can assist farmers with the banding, inoculation, trimming, and basic inspection of their flocks. Today your instructor will demonstrate techniques that you can use to handle chickens. These techniques must be done so that both you and the chicken remain calm and unstressed. Long gloves will be necessary for this handling experience to prevent scratches.

Reflective Questions

Answer the following questions relating to the activity.

1. Watch your instructor and your classmates. How do chickens behave as they are approached? What method works best for approaching them and picking them up?

____

____

____

2. Evaluate safety hazards that exist when working around poultry. What are hazards that a farmer should always remember to implement?

____

____

____

3. Create a testable hypothesis that can explain why chickens remain calm while being held upside-down.

____

____

____

____

____

____

____

____

____

____

Notes

Name ______________________ Date ____________ Period ________

CHAPTER 46 Poultry Selection and Evaluation

Carefully study the chapter and answer the following questions.

1. Why do birds molt? List four main causes.

 ______________________ ______________________

 ______________________ ______________________

2. Commercial layers are not raised with _____, so they only produce _____ eggs.

 ____

 ____

3. What are two reasons that genetic selection in poultry has been more influential than in other species of livestock?

 ____

 ____

4. What are the four classes of chicken breeds?

 ______________________ ______________________

 ______________________ ______________________

5. A normal eggshell can withstand about _____ pounds of pressure per square inch before cracking.

 ____

6. When comparing laying hens, a larger _____ is associated with a higher intensity of egg production.

 ____

7. The _____ of pullets and broiler breeder hens are monitored weekly to ensure good egg production and fertility.

 ____

8. Broilers and commercial layers are the same species. Why do they look completely different?

 ____

 ____

9. A Grade A carcass can have one or both _____ removed, but cannot be cut at the elbow joint.

 ____

10. The _____ can be seen on the outer surface of the yolk and can be used to determine fertilization status of the egg during breakout.

 ____

11. Grade B eggs can have a minor or moderate stain that covers less than _____/_____ of the shell.

 ____

 ____

12. During lay, _____, _____, _____, _____, _____, and _____ are regularly monitored to evaluate flock conditioning.

_______________ _______________

_______________ _______________

_______________ _______________

13. The _____ class has white ear lobes.

____

14. List three main objectives in the selection for improved efficiency in poultry.

_______________ _______________

____

15. The _____ is characterized by feather shanks.

____

Key Terms

Carefully study the chapter and then use the word bank to match the term to the correct definition.

A. air cell
B. albumen
C. check
D. cuticle
E. fleshing
F. Haugh unit
G. hen-day egg production
H. hen-housed egg production
I. molting
J. primary breeder

1. _____ A key measurement relating to the egg's protein quality and freshness
2. _____ The percentage of hens that are in lay at any time, accounting for flock mortality
3. _____ This develops on the large end of the egg between the two shell membranes
4. _____ The natural process of shedding and renewing feathers
5. _____ Sample genetic lines from poultry genetic companies that are crossed to produce parent lines to serve as breeders
6. _____ Ridges and grooves on the shells of eggs
7. _____ Action taken to assess proper body conformation for good egg production
8. _____ This makes up around 60% of the egg's weight and consists of four layers
9. _____ The percentage of hens that are in lay at any time
10. _____ A thin clear film deposited on the exterior of the shell

Critical Thinking

Using the information learned in this chapter, read the question and provide a detailed answer.

1. How can the FFA Poultry Evaluation Career Development Event be beneficial to participate in? What do participants learn that can be used in other areas of their lives?

____

____

____

____

Name ______________________ Date ____________ Period ________

Activity: Poultry Selection and Evaluation

Showing Poultry

Required Materials

- Laboratory notebook
- Pen or pencil
- Farm or laboratory outfit with gloves
- Poultry grooming toolkit
- Water for chickens
- Chicken

Activity Specification

Today your instructor will briefly teach you about aspects and attributes that a poultry judge looks for when evaluating show birds. Afterward, you will be asked to select a bird from your school's farm to show to your instructor and classmates. You will be expected to demonstrate proper handling procedures and discuss why you chose the bird you did.

Reflective Questions

Answer the following questions relating to the activity.

1. Analyze physical characteristics of poultry that are important for showmanship. Which three attributes did your bird show?

   ____

   ____

   ____

2. Evaluate the behavior of your bird while you are showing it. What aspect of showmanship could you improve?

   ____

   ____

   ____

3. Draw a chicken and label the following parts: comb, eye, ear, hackles, wing bow, wing bar, wing bay, breast, thigh, spur, shank, hock, fluff, main tail feathers, back, and any additional parts indicated by your instructor.

Name ______________________ Date ______________ Period __________

CHAPTER 47 Classification and Breeds of Poultry

Carefully study the chapter and answer the following questions.

1. The _____ breed is known for the striking color of its eggs, and does not carry a genetic defect that makes it susceptible to early death.
____

2. _____ are chickens raised for meat.
____

3. What bird requires a permit to be kept and bred in captivity, and can be found at poultry shows?
____

4. The _____ breed is very broody, and can be used as a foster mother to multiple poultry species.
____

5. _____ are typically processed for products like deli meat, sausage, tenderloins, and cutlets.
____

6. This game bird species is popular for hunting and raising in captivity. They are prolific layers and hearty meat birds, despite being smaller than other poultry species.
____

7. _____ are named after a Native American tribe indigenous to parts of upstate New York and Ontario, Canada.
____

8. This large breed lays many eggs, but this quality is overshadowed by its tendency to break them due to its weight.
____

9. Describe the differences between standard-bred, bantam, and variety breeds.
____
____
____

10. The _____ breed is responsible for almost every turkey and turkey product found on the shelves of American grocery stores.
____

11. The _____ breed was given to Queen Victoria as a gift and is known for its size; docile temperament; curved, fluffy tail; and feathered shanks.
____

12. The _____ of the Aylesbury duck distinguishes it from its almost-twin, the _____, which has a _____.

13. _____ are muscular birds that served as the cornerstone upon which the modern-day broiler was bred.

14. _____ are descendants of black Orpingtons that were crossed with local chickens after being imported from England in the late 1800s.

15. _____ do not make very good backyard egg layers but are extremely popular in commercial egg production flocks.

Key Terms

Carefully study the chapter and then use the word bank to match the term to the correct definition.

A. bantam
B. comb
C. exhibition type
D. plumage
E. production type
F. standard-bred
G. utility type

1. _____ Chickens that are further classified as general meat, egg-laying, or dual-purpose (meat and eggs) breeds
2. _____ The term for a bird's feathers as a collective whole
3. _____ Commercial meat or egg-laying chickens
4. _____ Smaller variety of a breed
5. _____ Regulates the chicken's body temperature
6. _____ Ornamental breeds
7. _____ The poultry equivalent of purebred

Critical Thinking

Using the information learned in this chapter, read the question and provide a detailed answer.

1. If you were responsible for creating a new crossbreed for commercial meat or egg production, which breed(s) would you choose? Why? What do you think this crossbreed would look like?

Name ______________________ Date ____________ Period ________

Activity: Classification and Breeds of Poultry

Presenting Poultry Breeds

Required Materials

- Computer with internet access
- Gmail account

Activity Specification

Today you will create a PowerPoint® or Google Slides™ presentation on poultry, while showcasing at least six different breeds.

Your presentation should:

- Include 10 slides total
  - Include a title page and a reference page
- Introduce the scientific name of the animal and any related scientific terms
- Describe the average age the birds reach sexual maturation, how they are bred, and length of gestation
- Present information about three breeds of poultry raised for meat and three breeds of poultry raised for eggs
  - Describe each breed's appearance, history, and purpose

Conclude with a five-minute presentation to the class where you present your slideshow and answer any questions.

Reflective Questions

Answer the following questions relating to the activity.

1. Analyze your classmates' presentations. What breeds were presented most frequently? Why do you think this is?

____

____

____

____

2. Evaluate your own presentation and indicate at least two places where you could have improved your communication.

____

____

____

____

____

____

3. Create a picture of what your ideal poultry breed would look like and describe three traits that would make this breed unique.

Name ______________________ Date ____________ Period ________

CHAPTER 48 Companion Animal Industry Overview

Carefully study the chapter and answer the following questions.

1. List the different segments within the companion animal industry.

___________________ ___________________

___________________ ___________________

___________________ ___________________

___________________ ___________________

2. Americans spend more than _____ per year to feed their pets.

____

3. Around _____% of global households have at least one pet.

____

4. What do adoption costs typically cover?

___________________ ___________________

___________________ ___________________

5. The largest segment of the support services industry is _____.

____

6. What are four questions one should ask to ensure a breeder is reputable before purchasing a companion animal?

___________________ ___________________

___________________ ___________________

7. Cats were domesticated about _____ BCE in the Fertile Crescent region, located in modern _____.

____

____

8. In the United States, _____ is the state with the highest percentage of pet owners, while _____ is the state with the lowest percentage.

____

____

9. There are more than _____ retail pet stores in the United States that sell live animals.

____

10. The most common categories of modern domesticated companion animals are:

____

____

______________________________ ______________________________

____

Key Terms

Carefully study the chapter and then use the word bank to match the term to the correct definition.

A. commercial breeder
B. companion animal
C. private breeder

1. _____ A person who maintains a small group of female companion animals for breeding
2. _____ Animal that lives, plays, and works with people
3. _____ Any breeder who keeps more than 30 females who are actively producing offspring

Critical Thinking

Using the information learned in this chapter, read the question and provide a detailed answer.

1. Imagine you are going to start your own companion animal breeding operation. Describe how you would ensure that you could be considered a reputable breeder and why each component is important. What would you need to get your operation started?

Name ______________________ Date ____________ Period __________

Activity: Companion Animal Industry

Veterinary Clinic Field Trip

Required Materials

- Laboratory notebook
- Pen or pencil
- Farm or laboratory outfit

Activity Specification

Today you will visit a local veterinary clinic. Make a careful observation of the layout of the facility and obtain other pertinent information regarding the office. Draw the veterinary clinic below and include the following aspects: All building entrances/exits, waiting room(s), patient room(s), imaging room, surgery room, fire extinguishers, AED machine.

Reflective Questions

Answer the following questions relating to the activity.

1. How would you describe the atmosphere in the veterinary clinic?

2. Evaluate the lighting, cleanliness, and organization of the veterinary clinic. Why is order important inside of a medical facility?

3. Create a list of three recommendations you have for improving daily operations and safety in the veterinary clinic.

Name ______________________ Date ______________ Period __________

CHAPTER 49 Companion Animal Management and Care

Carefully study the chapter and answer the following questions.

1. Parrots can live for _____ years.

____

2. Cats communicate with each other through _____.

____

3. Most at-home saltwater tank salinity is measured with a device called a _____.

____

4. Rodents can be especially sensitive to _____ and _____.

____

____

5. List four benefits of regularly grooming your dog.

______________________ ______________________

______________________ ______________________

6. Tropical fish need to be kept in water that is between _____ and _____ degrees Fahrenheit.

____

____

7. _____ prefer to live alone, while _____ prefer interaction and should have at least one other companion with them to keep them happy.

____

____

8. Cat gestation lasts approximately _____.

____

9. _____ _____ is a measurement used to explain the density of a substance in relation to the density of a standard substance.

____

____

10. The leading cause of infectious death in dogs is _____, causing death in up to _____% of dogs and _____% of puppies who become infected.

____

____

____

11. What illness can reptiles easily pass to people?

____

12. What are four indicators that your cat has internal parasites?

____

____

13. FOWLR (_____ _____ _____ _____ _____) aquariums include pieces of calcium with thousands of active _____ _____ on their surfaces.

____

____

____

____

____

____

____

14. A ball python's enclosure must remain at _____-_____% humidity.

____

____

15. The most common bird species kept as companion animals include:

____

____

____

Key Terms

Carefully study the chapter and then use the word bank to match the term to the correct definition.

A. fry	D. queen	G. terrarium	J. tortoise
B. insectivore	E. salinity	H. whelping	
C. parrot	F. tom	I. wing clipping	

1. _____ Animals that are similar to turtles but dwell completely on land
2. _____ The shortening of the primary flight feathers in pet birds
3. _____ The salt content of water
4. _____ Mature male cat
5. _____ A specialized type of carnivore that mainly consumes insects
6. _____ A broad description given to birds who stand upright, have clawed feet, and a characteristically strong, curved bill
7. _____ Glass enclosure with a screened top that is mostly used to keep reptiles
8. _____ Mature female cat
9. _____ Giving birth to puppies
10. _____ Baby fish

Name ____________________

Critical Thinking

Using the information learned in this chapter, read the question and provide a detailed answer.

1. Select a companion animal. Design housing, nutrition, health, and breeding plans specific to its needs.

Name ______________________ Date ______________ Period ________

Activity: Companion Animal Management and Care

Handling Companion Animals

Required Materials

- Laboratory notebook
- Pen or pencil
- Farm or laboratory outfit with gloves
- Pet nail clippers
- Colored pencil(s)
- Companion animal (cat or dog)

Activity Specification

Today your instructor will demonstrate how to safely handle cats or dogs. You will be expected to demonstrate proper handling skills. Then, you will practice your nail-trimming skills. You will use colored pencil tips as fake nails, as they have a similar consistency to animal nails.

Reflective Questions

Answer the following questions relating to the activity.

1. Analyze the behavior of the companion animal as it was handled. What expected and unexpected behaviors did you observe?

____

____

____

____

____

____

2. Evaluate the consistency of the colored pencil tip that you trimmed. How could your trimming be improved?

____

____

Name ______________________________

3. Create an image sequence below that demonstrates what an unclipped and clipped pet nail might look like.

Before clipping:	After clipping:

Notes

Name ______________________ Date ____________ Period ________

CHAPTER 50 Classification and Breeds of Companion Animals

Carefully study the chapter and answer the following questions.

1. The _____ dog breed comes in several shades of gray, designated as blue, silver, and silver gray, and has light-colored eyes.

____

2. The Persian cat breed has a body type described as _____, meaning it has a short neck and legs.

____

3. One of the most popular rabbit breeds, _____ rabbits are known for their distinct white collars.

____

4. This medium-large cat breed is prized for its intelligence, with many owners noting they are more like dogs than cats.

____

5. _____ is the French word for *low*.

____

6. _____ are members of the weasel family, with extremely long bodies and short legs.

____

7. _____ were originally bred to use their vision to track and locate prey, and should have large feet; prominent hipbones; long, springing gaits; and regal or aloof expressions.

____

8. _____ rabbits are a relatively new breed and have ultra-soft, plushy fur.

____

9. _____ were bred to serve as work dogs in fishing operations, and were used to herd fish into nets, retrieve fishing gear, and swim between boats and shore.

____

10. _____ are stocky felines with short hair and ears that are rounded at the tips.

____

11. The _____ is often recognized as the most intelligent dog breed, and originated in the hills between Scotland and Wales.

____

12. _____ are nocturnal, which is why they have such large eyes; they get their name from their desire for sugary foods like sap and nectar.

____

13. The _____ cat is recognizable for its silver-blue coat and green eyes.

____

14. The "badger dog" has evolved with courage that is much larger than its small stature. What is this breed?

____

15. List two notable dog breeds that originated in Scotland.

____

____

Key Terms

Carefully study the chapter and then use the word bank to match the term to the correct definition.

A. brindle	D. merle	G. ticked
B. herding group	E. sporting group	H. toy group
C. hound group	F. terrier group	I. working group

1. _____ Small dog breeds, including those that are naturally small and those that have been bred as smaller versions of their larger counterparts
2. _____ Dog breeds developed primarily as hunting companions, chasing down warm-blooded game
3. _____ Dog breeds that were originally bred to assist humans in hunting and retrieving birds
4. _____ Coat pattern with mottled patches or shades, often with white intermixed
5. _____ Dog breeds that were bred to aid their human companions
6. _____ Dogs who were originally bred to gather, herd, and protect livestock
7. _____ Dog breeds that were primarily bred to hunt vermin and other small prey and to guard homes and barns
8. _____ Coat pattern characterized by dark specks of color on a white background
9. _____ Coat pattern with streaks or stripes of color

Critical Thinking

Using the information learned in this chapter, read the question and provide a detailed answer.

1. Imagine you are going to get a companion animal. What type and breed of animal would you select? Why? Be sure to discuss temperament, budget, and other considerations you would make.

Name ______________________ Date ____________ Period ________

Activity: Classification and Breeds of Companion Animals

Preparing a Rabbit for Show

Required Materials

- Laboratory notebook
- Pen or pencil
- Farm or laboratory outfit with gloves
- Rabbit
- Rabbit-grooming toolkit
- Water for rabbits

Activity Specification

Today your instructor will briefly teach you about the aspects and attributes that a rabbit showmanship judge looks for when evaluating rabbits. Next, your instructor will teach you about grooming and handling a rabbit. Then, you will be asked to practice showing a rabbit in small groups, demonstrating proper handling skills.

Reflective Questions

Answer the following questions relating to the activity.

1. Analyze the behavior of rabbits and describe two safety problems that could result from handling them.

____

____

____

____

____

____

2. Evaluate the physical structure of your group's rabbit. In what areas is your animal lacking desired attributes? What areas exceed desired expectations?

____

____

____

____

____

____

3. Create an infographic designed to educate others about proper handling and safety concerns that may arise when working with rabbits.